Starting A Home Workshop

CW00631719

Starting A Home Workshop

ANDREW SMITH

Model and Allied Publication, Argus Books Limited

Model & Allied Publication
Argus Books Ltd.
Argus House
14 St James Road
Watford
Herts
England

© Argus Books Ltd 1980

© Andrew Smith 1980

ISBN 0 85242 725 5

All rights reserved. No part of this book may be reproduced in any form or by any means without written permission of the Publisher.

Cover Photograph by kind permission of
N. Mole Ltd. Watford

Typeset by Inforum Ltd. Portsmouth
Designed by Kaye Bellman
Printed and bound by
W & J Mackay Ltd. Chatham

Contents

Introduction

A pair of hands and a few tools are the prerequisites but it is the place where the work is done that makes the "workshop".

Like so many things in life, we may find that the work-place tends to grow on us haphazardly as the jobs multiply and our interest in craft-work increases, until finally we decide that something must be done about it. The kitchen table is just not the place to build that new bookcase, and the back of an ordinary garage is a miserable site in which to build a model steam engine when the temperature is below freezing. Likewise there have been distinct mutterings about "bits in the carpet" since we became interested in ship modelling and clamped that small vice to the window sill in the bedroom!

In other words, if you have ever felt the need for "a place of your own" where you can saw, drill, file and build to your heart's content, then perhaps the ideas in the following pages may help you achieve it. Even though it may be only a corner, or a tiny shed in the garden, it will repay careful thought and planning so that you may work there comfortably and efficiently.

Although all the ideas and suggestions given in these pages are the result of experience, do not just copy them without giving the scheme very careful consideration. No two workshops are alike, and similarly, someone else's workshop will not exactly suit you or your particular interests. In a similar vein, do not apply other people's "rules" to your workshop. The adage, "a place for everything and everything in its place", is an excellent one if you are a tidy person, but it could mean that nothing ever gets used and therefore nothing is ever made.

In my own workshop many of the commonly used tools live on top of the bench, in fact, any new tools I buy generally reside there until I decide they are in the way and should be found a "home".

We have given a fairly detailed account of the "whys" and "wherefores" of obtaining permission either to adapt an existing location for use as a workshop, or to erect a small building in the garden. Our purpose has been to show that the procedure is perfectly simple and straightforward.

A phone call, letter or visit to your local council offices will put you on the right track with the minimum of fuss, and, who knows, they may even suggest something you hadn't thought of.

If reading these pages has introduced you to the joys of owning a workshop, you will find, in the wide range of M.A.P. publications, a magazine devoted to the enjoyment of your new hobby; while the Argus Book List contains many books aimed at helping to develop your expertise and skill.

These remarks would not be complete without giving my thanks to the two expert contributors: Eric Woodward, lecturer in electrical installation, who dealt with the technicalities of getting electric power into an outdoor workshop, and Andrew James, solicitor with a large County Council, who has helped to guide us through the problems of getting permission to build a workshop.

Finally, and most important, the reader.

This book is for the real newcomer to owning a workshop, in fact one might almost say, the pre-beginner, about to take the plunge. Hopefully it will answer some of his questions and put him on the road to developing a satisfying and worthwhile workshop hobby.

Saltford
Avon, England.

Andrew Smith

Acknowledgements

Sincere thanks are due to the many firms and suppliers who, so willingly, offered illustrations and helpful suggestions during the preparation of the text, and who put up with my pestering as publication date drew near.

Among many others, they include:

Marley Buildings Ltd.
Silver Mist Ltd.
Briklap Ltd.
Europleasure Gas Ltd.
Lervad Benches Ltd.
Burgess Power Tools Ltd.
Record Ridgway Tools Ltd.
Perfecto Engineering Ltd.
Aven Tools Ltd.
Stanley Tools Ltd.
Neill Tools Ltd.
Henley Engineering Publications Ltd.
Elliott Machine Equipment Ltd.
Myford Lathes Ltd.
Cowells Ltd.
Avon Models.
Black & Decker Ltd.
Parwin Heaters Ltd.
GB Shelving Ltd.
Woking Precision Models Ltd.
W.A. Meyer Ltd.
Link Plastics Ltd.
Compton Buildings Ltd.

And a special thank you to Kaye Bellman, Argus Ltd., book designer, for her unceasing efforts to give character and style to the amorphous mass of typescript, diagrams and illustrations that formed the original manuscript.

Andrew Smith

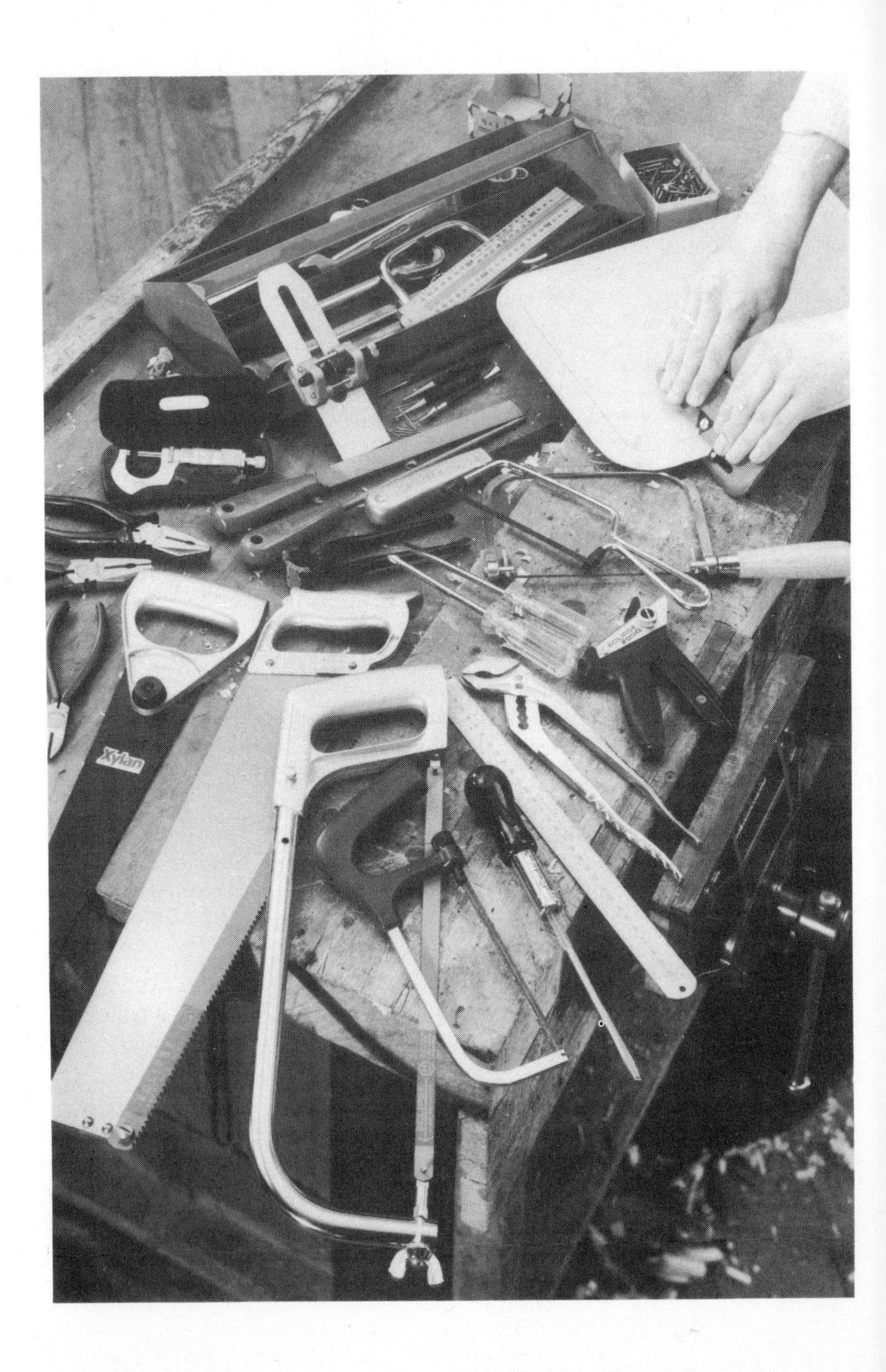
Xylan

1. WHY HAVE A WORKSHOP?

The simple explanation is that a workshop is a place where we can make and repair things. But it can also be a little haven of peace where we can shut out this mad, turbulent world and evolve a 'make believe' environment where the hiss of a shaving emerging from the throat of a well sharpened plane and the feel of an accurately turned piece of brass gives us such pleasure as is difficult to understand and embarrassing to explain.

The desire to create is in us all, but it is only when we have a workshop that we can give full expression to this need.

Why do we have the desire to make things?

The characteristic that raises the human race above the animals is its creativity. Early man's realisation that his hands were tools with which he could fashion various artefacts to help make life more bearable and less of a mere struggle for existence, and the subsequent experience that caused him to develop tools that were, in effect, extensions of his own hands, were the factors that gave him the control he now has over natural materials and effects.

As life became more sophisticated, the need for each individual to create for his own personal comfort became less necessary and has been replaced by an economic and social situation where most of our requirements, from food to furniture, are obtained by simple purchase.

Fig. 1/1 A few tools and a pair of hands are the prerequisites, but it is the place where the work is done that makes the "workshop".

Although this would appear to be an ideal life style, it has, in fact, caused the individual's desire to create to be largely unfulfilled. And so we turn to making things, not because they are needed for living, but because the act of making them and the satisfaction we derive from using (and perhaps playing with) them gives us a wonderful feeling of pleasure and contentment.

This is true at any age, and regardless of whether the result of our efforts is a young kid's pram-wheeled trolley or a grown-up kid's super-scale model steam locomotive. Quite simply, the physical and mental processes involved in building something, however simple, makes you and me better persons, both in ourselves, and in our relations with all the other people with whom we come in contact.

What kind of workshop?

There is no such thing as the 'perfect' workshop, just as there is no such thing as the 'complete' workshop. As long as you find happiness in your workshop it will continue to grow and develop. I have had a workshop of one sort or another for forty years. I still use tools bought all those years ago, yet I have recently improved it further by adding arc-welding facilities, a development I would never have even remotely imagined when, as a young teenager, I first turned the garden shed into a workshop.

Making a start

We have been informed from what, in journalistic circles, would be described as "a reliable source", that old man Confucius, that wily Chinese sage, once remarked, "journey of thousand miles start with first step"!

Now personally, I am somewhat doubtful whether the revered gent really did make that remark, because with his sagacity he would certainly have foreseen the advantages of quoting the metric units, and with his obviously elevated position in the society of the day, he would have first made sure of a comfortable litter!

However, the axiom is a good one when we come

Fig. 1/2 A large model stationary steam engine standing on the lathe on which it was built, in a garden workshop.

to start a workshop. By all means dream about that warm, comfortable workshop with a thousand pounds' worth of equipment, but don't sit on your backside to dream; dream while you search out all the old tools and bits and pieces that are scattered about in the house, shed and garage.

All tools are valuable, they are worth keeping, reconditioning and looking after. That old screwdriver with its handle split, was bought in Woolworth's for 6d ($2\frac{1}{2}$p to our younger readers), donkey's years ago; it would probably cost you a pound today. So get some good quality wood glue, open up and clean the split in the handle and glue and clamp it together. After setting, carefully file the screwdriver edge straight and flat so that it will remove or drive home a screw efficiently. You may then be surprised at how well it does its job.

The odd woodwork chisel may come to light, perhaps sharpened so often in the past that it is now only a couple of inches long. Never mind, it is worth refurbishing. Later we will tell you how to sharpen the chisel *and* that old saw that has been used for hacking overhanging branches from the neighbour's tree!

An old file, full of paint, can be given a new lease of life by being cleaned with a sharp pointed wire if a wire brush or piece of file card wire is not available. But first, buy a file handle and fit it. It will only cost a few pence, even in these days, and safety must always be our main theme – I've seen a poor chap with a file tang in the palm of his hand, not a pretty sight.

Gather all the tools together in one place and have a really good look at them, perhaps for the very first time! A word to close relatives and friends may bring other items your way, unless of course, they are also reading this!

What else is there?

Well, materials come next. Gather up all that might be useful. Sort out any timber, but throw out for burning or for the local council waste dump, all that is infested with beetle or has decay. Although if you are a gardener, it might have enough life left in it to act as a plant stake if treated with some preservative.

All metal should be kept. Even rusty steel can be cleaned and found useful. Your attitude should be, that before you throw anything away, ask yourself how much it would cost to buy such a piece of metal, wood or plastic, if you wanted it for a job. Remember, rubbish is only rubbish when it is in the way; when it has been sorted and stacked it becomes an invaluable store of new material! One of the most useful items in the workshop of the enthusiast is the scrapbox.

The same goes for nuts, bolts, screws and nails. Sort them into types and sizes and store them in labelled tins or jars. There was a time when one might be laughed at for such penny-pinching ways but not these days. If you get hold of any bits of equipment, old radios, machinery parts, etc., strip them, sort and store the bits as if it were gold – it nearly is! Their worth is twofold; first you will learn a great deal about how things are made and work, and second you will increase your treasure trove of bits and pieces. For example, should you want a few 6 BA screws for a modelling project, you will find old electrical and radio equipment full of them, but if you wanted to buy a few, you would probably be very surprised at what they would cost.

A place to work

Wonderful examples of craftsmanship have been made on a corner of the kitchen table, and I remember, in my own very young days, when fretwork was popular, making do with a light, baize-topped folding card table. I had to stop the table from walking away as well as trying to shape the fretwood. When I progressed to model aircraft, the table surface

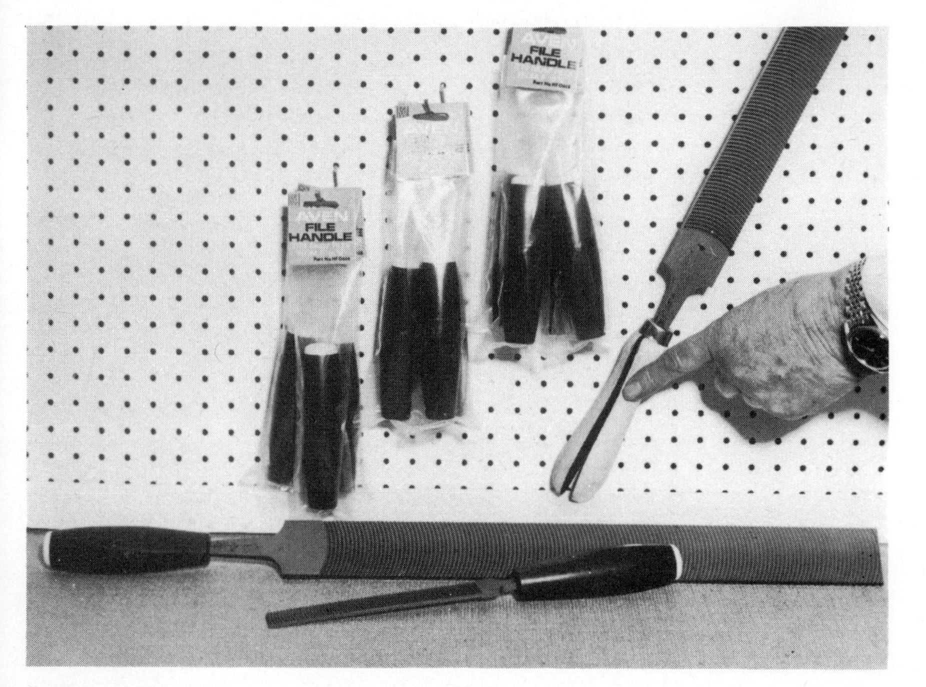

Fig. 1/3 Never use a file with the handle in this condition.

became more and more sparse, like a lawn during the summer of '76 as the baize got pulled away, stuck by balsa cement to the aircraft's structure.

The most important feature of a workplace is to be able to leave work and tools out. It is nice not to have to be tidy! If you can annexe an inside room for a workshop you are very lucky. Basements have great possibilities, attics rather less so. The back of the garage may suggest itself but like the garden shed, needs a bit of attention, unless we intend being a warm weather "workshopper" only. However "beggers can't be choosers", as I believe Confucius said at the end of his thousand mile journey, so get what you may and between us we will do all we can to help you fit it up as a pleasant and comfortable workshop. One day we may even have the pleasure of visiting you there.

The question of cost will loom large in your planning and decisions. And while it must seem wonderful to be able to purchase all the equipment and materials you desire without giving thought to the

Fig. 1/4 Safety Goggles.

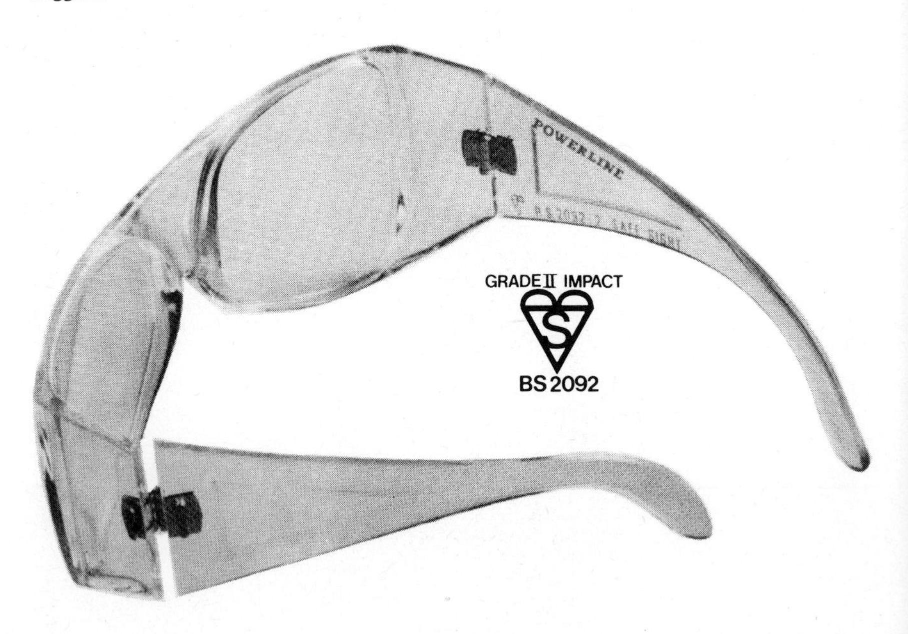

Fig. 1/5 Safety is all important in any workshop, especially of the eyes. Wear SAFETY GOGGLES at all times, make it a habit.

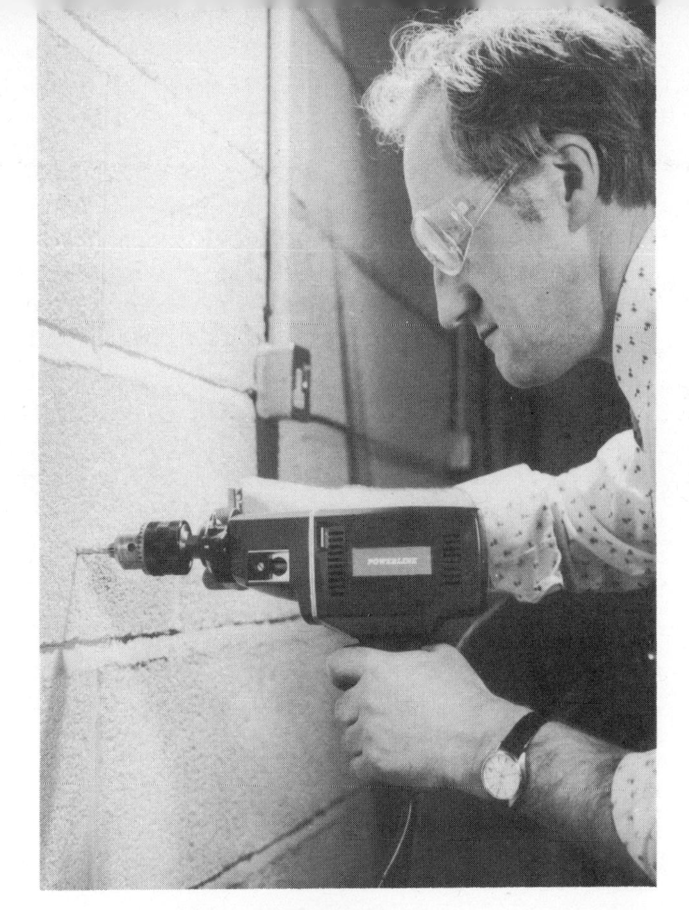

Fig. 1/6 Wear safety goggles when driving masonry nails. These nails are hard and may splinter.

state of the bank balance, nevertheless a great deal of the pleasure of a workshop is in keeping its expense down to a minimum.

After all, the reason we desire such an establishment is so that we may make rather than buy and, to quote Granny, if we can "make silk purses out of sows' ears", the exercise is all the more satisfying.

Safety

Before proceeding further it must be emphasised that SAFETY is of paramount importance in all workshop activities and although reference will be made to this frequently, it is something that is entirely in your hands. No one can force you to work safely and maintain a safe environment. Even in an industrial situation the efforts of Safety Officers and Factory Inspectors can be of little avail if the individual chooses to work in an unsafe manner. Hence working safely is a personal discipline – cultivate it.

2. WHERE TO BUILD A WORKSHOP

Having decided that we want a workshop, the next step is to plan its location and first let us consider the possibilities that may exist within the structure of the house itself.

In the house

If we can purloin a spare room we really are in luck. Perhaps it is a bedroom once used by a son or daughter now grown up, married and with a home of their own, or it may be a ground floor room in the older type of house, warm, dry and comfortable. There will be little or nothing we need do of a structural nature to make the accommodation suitable, but I think it is worth giving some consideration to the effect of an indoor workshop on the rest of the house.

The floor needs protecting, especially if our hobby entails the machining of metal. Swarf and oil will play havoc with a wooden floor; the room may well revert to its more normal purpose, and if you move, the new owners will not appreciate floor boards soaked with oil and embedded with steel swarf. Sheets of hardboard are very cheap and form an excellent protective covering. For a few pounds it is worth covering the whole floor before any other equipment is moved into the room.

While on this topic, it is a good idea to keep an old pair of shoes to wear in the workshop. These never leave the confines of the room, being kept, when not

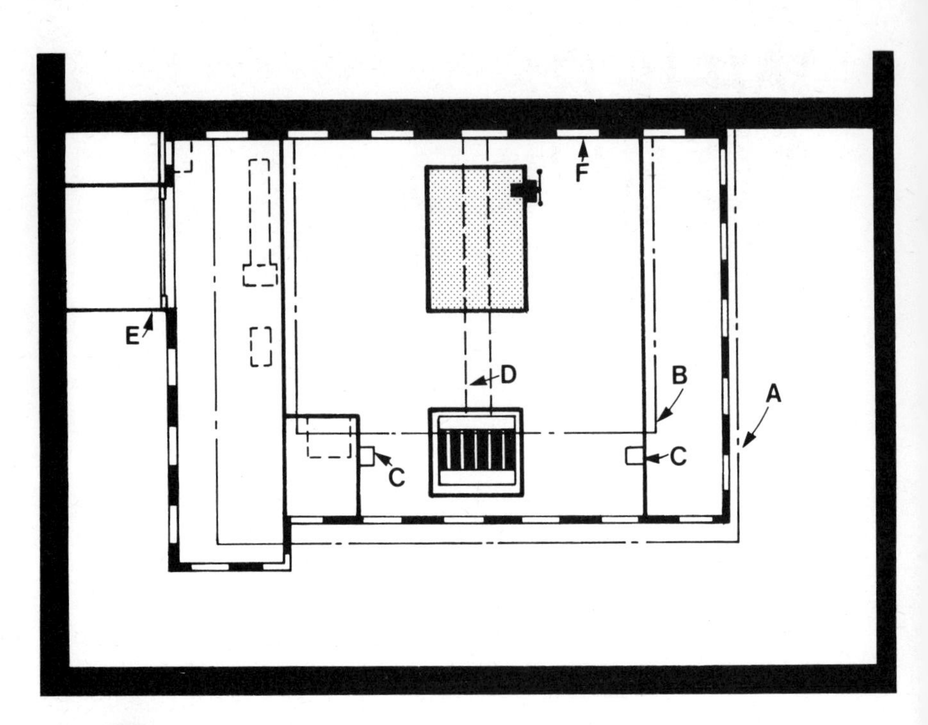
F
E
D
B
A
C
C

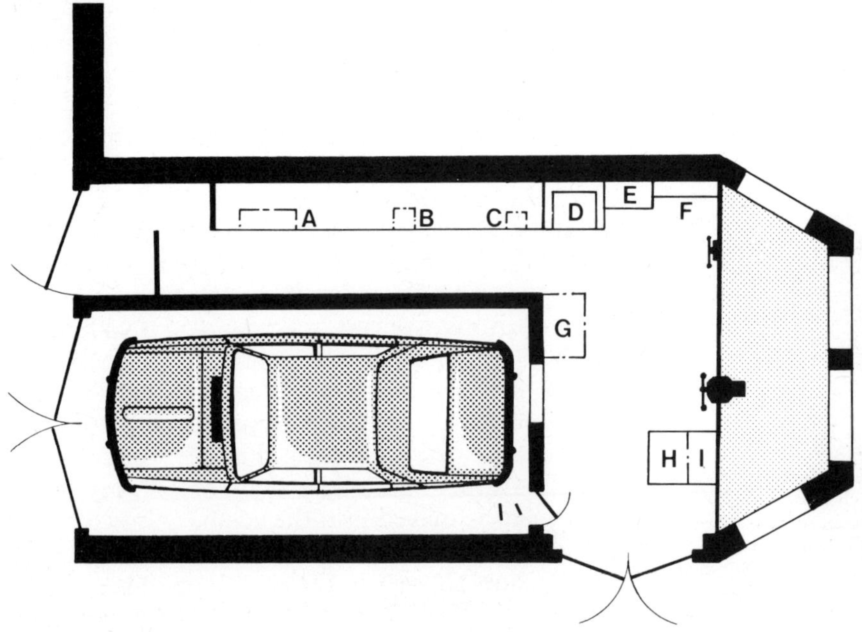
A
B
C
D
E
F
G
H
I

Fig. 2/1 Plan of an attic workshop. Principal issues as indicated are, 'A' 4 ft. height line, 'B' 6 ft. height line, 'C' section of roof construction, 'D' dotted line showing position of parting wall below, 'E' position of a window, 'F' essential sound-absorbing panel.

in use, just inside the door so that there is no risk of transferring sawdust or swarf about the house. The latter plays havoc with carpets.

Wall protection is needed where there is a risk of oil being thrown around by a lathe or drilling machine. On the wall, oil will always stain any subsequent redecorating that may be carried out. Here again, hardboard may be used as splash backs for the lathe and drill bench. If you, instead, put some protection directly on the wall, make sure that it cannot, in time, be penetrated by the oil; aluminium baking foil is excellent stuff to use, held on with drafting tape or similar.

Noise may be a nuisance, although the intermittent sound of benchwork is often less annoying than the constant rumble of a lathe. If, to increase their rigidity, we fix benches and machine tool stands to walls and floors we will probably find that the noise is transmitted by the fabric of the house. If this proves to be troublesome, the only solution is to isolate our workshop equipment by fitting rubber pads under the legs of machine stands. This necessitates that such stands be made sufficiently robust in the first place, which is a good thing. Good, stout timber

Fig. 2/2 Garage workshop – principal items are, 'A' Lathe, 'B' power drill, 'C' electric bench grinder, 'F' paint and tool storage, 'G' milling machine, 'H' circular saw and planer, 'I' bandsaw.

Fig. 2/3 Whichever type of workshop is chosen, a substantial concrete base is a necessity. Its top surface should come well above the surrounding soil level.

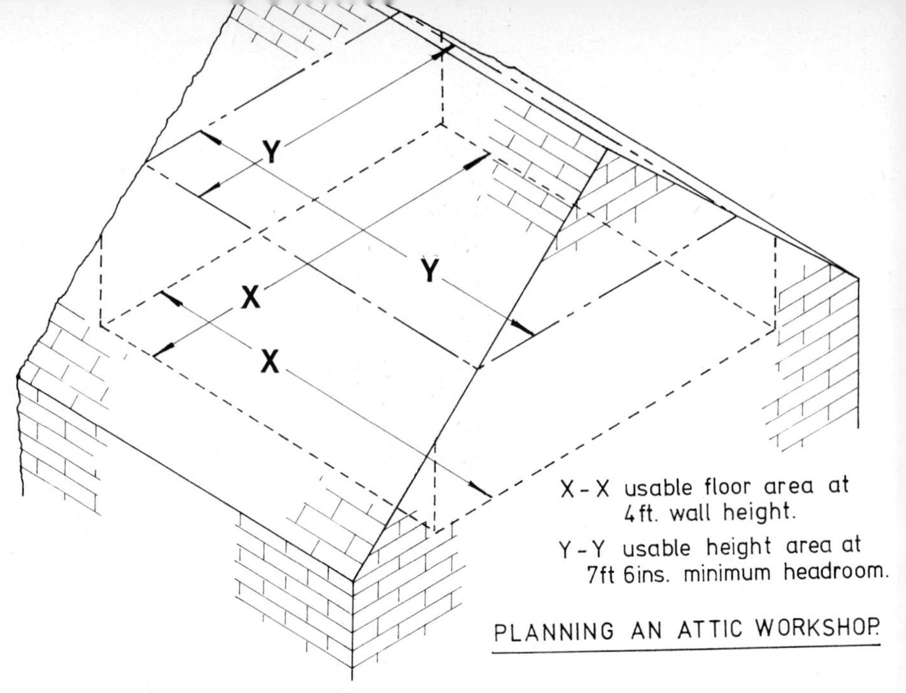

Fig. 2/4 The initial planning of an attic workshop may be based on the information shown. You may feel able to make do with less than the 7 ft 6 in headroom, but it is a normal building regulations requirement.

benches seem to have a much better damping capacity than those made of metal. Remember that noise is due to vibration so ensure that machines and motors, vices, etc., are all firmly bolted in place so that, as far as possible, they cannot vibrate.

If the noise is still troublesome and you cannot pull rank and tell the family not to notice it, or if the room has a wall shared with neighbours, you may consider lining it with a sound insulating board. Alternatively, hanging a heavy curtain on the wall will considerably reduce any sound passing though it.

Attics and basements

Parts of a house which may seem to be an obvious choice for a workshop without encroaching on the more normally used rooms are the attic and the basement. Problems of access may limit the usefulness of the former, but if the roof permits sufficient headroom the space is certainly worth considering. The difficulties are noise, which has already been commented upon, and, more serious, floor support. It must be remembered that we are using, as the support for our workshop floor, the ceiling joists of the room below and these are unlikely to be as substantial as floor joists. The answer is to locate our attic workshop in such a position that we have as much support as possible from the walls of the rooms below.

The joists must be floored and for this I advise sheets of chipboard at least $\frac{3}{4}$ inch (20 mm) thick, screwed down after all wiring and pipework has been carefully and neatly let into the joists. At this juncture it is advisable to make a drawing of the position of all pipes and wires which lie beneath the floor boards to make any subsequent repairs easier. Because the flooring will no doubt have to come into the attic via the access hatch, it will need to be cut into suitable sizes; try and make this to suit the centre distance spacing of the joists.

Fig. 2/5 An 8 feet by 6 feet garden shed type workshop from Silver Mist Ltd. This is a type widely used by many enthusiasts and if lined to keep in the warmth and regularly creosoted to maintain the timber, will last for a very long time. It is advisable to mount it on a concrete base raised a few inches above the surrounding soil level.

Walls and ceiling for the attic room may be lined with a fibre building board on 2 inch (50 mm) square supports. An opening roof light is worth considering, not so much for light, as most of our use of the room will be in the evening, but to admit fresh air into the workshop.

Basements come in various shapes and sizes, from the semi-basement with windows high in the wall to real cellars or others which are only glorified coal-holes. If dampness is a problem, the walls may be treated with proprietary sealers, while the floor, if flagged or concrete, may have a damp-proof sealer applied, surmounted by a new floor of wood blocks or chipboard. This will make it considerably warmer underfoot.

Fig. 2/6 An excellent type of outdoor workshop, the Marley "Monarch". Available in a range of sizes and made from thick pre-cast concrete panels. The panels incorporate a tongue and groove joint mastic sealed during erection which effectively weatherproofs the structure. The incorporation of lining clips during assembly to take a dry interior lining is very worthwhile.

Fig. 2/7 For the enthusiast to whom the workshop is becoming more than just a hobby, the Marley "Midhurst" range can offer the possibility of having separate facilities for a drawing office or welding and heat-treatment. Did someone suggest "dog-house" facilities?

Many of the older type of terraced town house have basement areas in the front reached by a stone or iron staircase from the pavement. Part of this area may be roofed over using corrugated plastic sheet to make a workshop. With the walls treated with a white cement or similar type of paint a useful work-room can result. It has the advantage that with access directly from the street one can smuggle in lengths of metal, bits of machinery, etc., without the necessity of explanations to the domestic boss!

The garage workshop

As much of the maintenance and repair work we carry out is on the car, it may seem sensible to develop our workshop in the back of the garage. Where it is built as part of the house and can gain some warmth thereby, the idea is a good one. The chief fault arises if the garage happens to be minimal and the car large leaving little room for benches, let alone our bulk. To work on a frosty winter's night with the doors wide open and the car standing half out of the garage is not my idea of pleasure! However with a growing preference for smaller cars and if we

Fig. 2/8 Briklap construction, the lower part of this single storey building shows the normal Briklap finish; the upper half of the wall is the tile finish brick.

park really close to the side wall and just inside the door, we may find that there is sufficient room to arrange a cosy little work-den. If we can put up a partition to separate the workshop from the rest of the garage, things will be even cosier. Even a heavy curtain or old carpet hung from the roof will make a difference.

Should you happen to be in the market for a garage and if you are considering purchasing one of the sectional, pre-fabricated type, it is worth considering extending it to form the desired workshop. The price of an additional section or two will be nowhere near as expensive as a separate workshop while the rates will be for a single building rather than two. An internal partition to keep the workshop section separate and cosy may easily be made from timber studding covered with building board. Even the luxury of a communicating door is not difficult to fix into the partition.

Garden shed workshop

This, I imagine, will be the most likely form of workshop used by readers and, in my opinion, has distinct advantages as long as one is prepared to develop it into a pleasant and comfortable workplace and not let it degenerate into a repository for garden tools and cast out items no longer welcome in the home. Consider the advantages. It is isolated from the house and noises, dirt and fumes will not be noticed. The structure and fabric may be altered in any way to suit its purpose as a workshop. It can be locked and isolated and will not be intruded upon by other members of the family – particularly important if there are young children, to whom a workshop is a source of danger.

This kind of workshop will probably be timber built, but a more permanent brick or block structure is not difficult to construct, while the use of galvanised steel or aluminium sheet should not be overlooked, especially if your skills tend to favour working in metal. Sectional buildings are available in all three forms of construction and materials.

In your choice of such a building, it is worth looking for one that is fitted with clips so that an inner lining may be fitted. This is obviously easy if the structure is of timber but less conveniently arranged if metal or a pre-cast concrete.

The first shed I had as a workshop was one my father built. It consisted of a 2 inch by 2 inch timber frame-work covered with galvanised steel sheets obtained by opening out old forty gallon drums which had held some commodity or other. The inside was lined with asbestos cement sheets.

Whichever type is chosen, a good concrete base is a necessity with its surface well above the level of the surrounding soil. The floor of the timber workshop will then rest on this, supported on its joists, while the other two may have a chipboard floor laid, well bedded in a waterproof sealer.

Fold-away workshop

The flat dweller, single person in a bed-sitter, or perhaps the occupier of council Senior Citizen accommodation need not feel that workshop facilities are beyond them. There are many examples of workshops built into recesses, in roll-top desks, kitchen cabinets, wardrobes, etc. A "Ham" friend of mine has a very complete radio station in a built-in wardrobe in his bedroom.

The principal factor is to accept that the size of work attempted should bear a relationship to the space available. If your interest is live-steam locomotives, then choose $2\frac{1}{2}$ inch gauge examples rather than $7\frac{1}{4}$ inch gauge. If woodwork, concentrate on small examples of fine cabinetmaking or carving and not large wardrobes and dining tables. The pleasure of creating depends in no way on the size of the article. The enthusiastic plastic modeller may find that a large wooden tray across the arms of his chair, together with a small side-table as an overflow, is all he needs for many happy hours.

Bricklaying made easy

With labour costs so high, there will be few readers who will not give some thought to building their own workshop. The choice of material then becomes the important factor and if brickwork is necessary, perhaps to match existing buildings, there is now a brick available that makes the laying a simple and speedy job for the complete amateur.

BRIKLAP is a specially shaped brick now being used in the construction industry, with the mortar joint concealed by the brick's shape. No tedious pointing is required and the need for building lines, trowels and other equipment is much reduced.

Conventional foundations are prepared in the usual way and the bottom course is laid as normal and carefully levelled. From then on the mortar, mixed to a creamy consistency, is poured into the special plastic gauge which is pushed along the wall leaving a trail behind of the exact quantity of mortar.

The perpendicular joints are automatically filled as the gauge passes over and the next brick is just slid into final position.

Briklap has been used in the construction of single and two-storey buildings and is suitable for single,

Fig. 2/9

1.

2.

3.

4.

5.

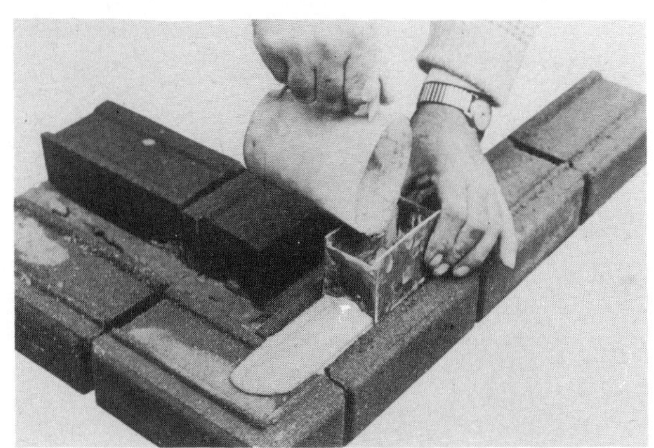

6.

7.

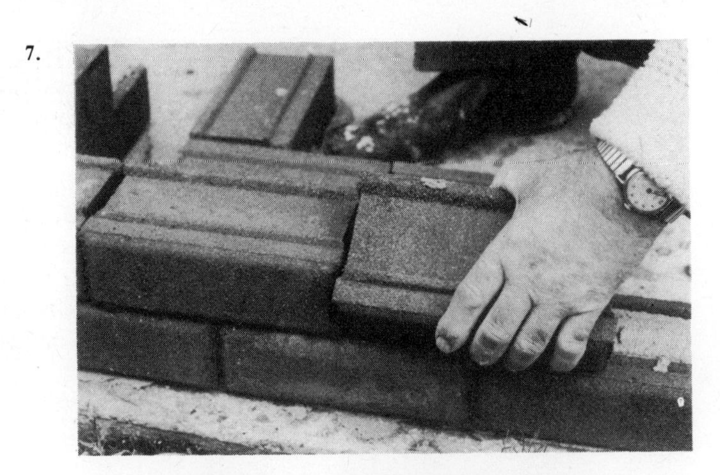

8.

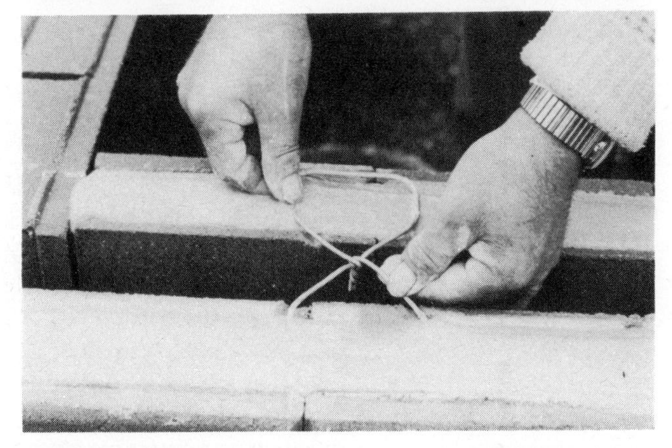

9.

double and cavity wall construction. In addition, a wide range of brick styles and finishes are available to add visual interest to the building scheme.

Because the system may be new to many it is worth having a look at the technique that would be employed by the amateur in using these accurately moulded concrete bricks; this is explicitly shown in the illustrations 1 to 9 of Fig 2.9.

1. On a conventionally prepared foundation the first course (layer) of bricks is bedded into a thick layer of mortar.

2. Great care is taken when laying the initial course to get it flat and true.

3. Use a square and spirit level to ensure accuracy.

4. If a cavity wall is required, the inner skin is started on a thick mortar bed and positioned to leave the required width of cavity.

5. Subsequent courses are laid using the mortar gauge. The mortar, mixed to the consistency of thick cream is poured into the gauge as it is slid along the brick guides. A mortar layer of the correct thickness is left behind; mortar also flows down between the bricks to lock them together.

6. Ensure that corner bricks are 'plumb' on both vertical faces by checking with a spirit level.

7. Each brick is placed in position with a slight gap and then slid along firmly into position, ensuring a precise joint.

8. Wall ties to bind together the inner and outer skins of a cavity wall are fitted by chipping a little locating groove in the bricks, laying the tie in place and then running in the mortar.

9. Bricks with a decorative face may have their vertical accuracy checked by plumbing from inside the wall cavity.

3. LIGHT, HEAT & POWER

The alternative to daylight

As most of our use of the workshop is likely to be in the evening, it is unlikely that we can make as much use of daylight as we might like to.

One of the rules of the early craft guilds was that members were not permitted to work by artificial light, i.e. candle or oil lamp. It was considered that the quality of craftsmanship would suffer if the work were not done by daylight.

Regretfully in our own amateur workshops most of our craftwork will be performed by the aid of artifical light, so in deference to the ancient guilds and, more important, for the benefit of our eyes, let us ensure that the quality of lighting is as good as possible.

The problems discussed in this chapter will not concern us greatly if we have an indoor workshop, but if we have a garden shed type of establishment they will be important, and the further it is from the house the more difficult will the problems of light, heat and power become.

Mains electricity

If the workshop is sufficiently close so that we may run an electrical supply from the house, much of our problems will be solved, and a later article, by a qualified lecturer in electrical installation practice, tells us how to set about this. For the present let us consider the important details of what form of heating a workshop fitted with mains electricity may employ.

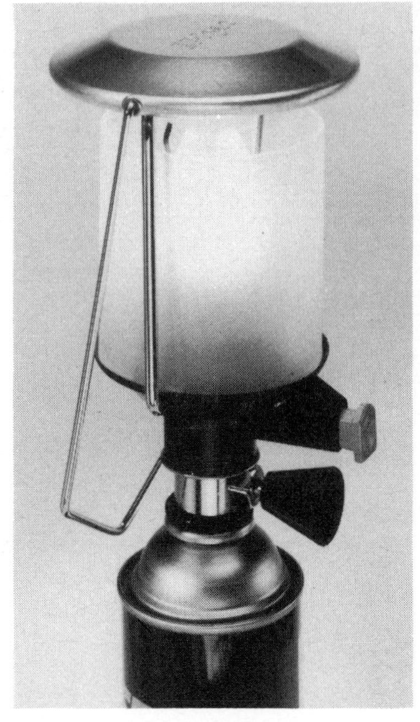

Fig. 3/1 The EPIgas heater is of approximately 750 watts output; it is lightweight, portable, economical and effective. It will allow you to warm your workshop on chilly days. The sturdy carrying handle and guard conform to British Fireguard Act standards.

Fig. 3/2 With a light equivalent to approximately 80 watts, this type of L.P. Gas Lantern is ideal for use in small workshops where electricity is not available.

The easy choice is, of course, an electric fire. This is a convenient, efficient and rapid heat source. But any situation in a workshop where we have naked flames or glowing elements is always a source of danger. With sawdust and shavings lying about or oil soaked rags, there is always the risk of a fire. For this reason it is advisable to find a method of heating which has no exposed high temperature source.

Two possibilities are available. If we use the workshop at infrequent intervals we obviously want something that will quickly remove the chill from the air and make things reasonably comfortable. For this, the blower type of heater as sold for greenhouse heating is ideal. Carry out a few experiments to find the best location for the heater so that the warm air will circulate round and round the workshop, bearing in mind that hot air rises and so conversely, the coldest air in the shop will be near the floor. Also don't have the warm air blowing at your face especially your eyes; this is most uncomfortable.

The second possibility, and the one I use, is an off-peak or night storage heater. I use my workshop fairly frequently and, to me, the idea of maintaining it at a steady (but not high) temperature, seems the right thing to do. The heater I have is of the metal cased industrial or office type, such as one would not normally have in the home and cost me £5 secondhand. It is of 3 kilowatt output, but is run at its lowest setting and I find it only needs to be switched on during alternate nights, in the coldest weather. On entering the workshop from the house, it feels cool, but a boiler suit over my normal clothes and some hacksawing or filing at the first job soon gets me warm enough to spend the rest of the evening standing at the lathe.

Of course the necessary wiring from a special off-peak supply is needed but you can do this apart from the final linking in to the mains supply and the fitting of the special meter, etc. Details of all this will be available from your local electricity board, and secondhand heaters are frequently advertised in local papers. If the workshop has frequent use it is well worth considering.

'Gas' lighting

In cases where it is inconvenient to obtain a mains electricity supply to a garden workshop, other means of supplying light and heat may be considered. A

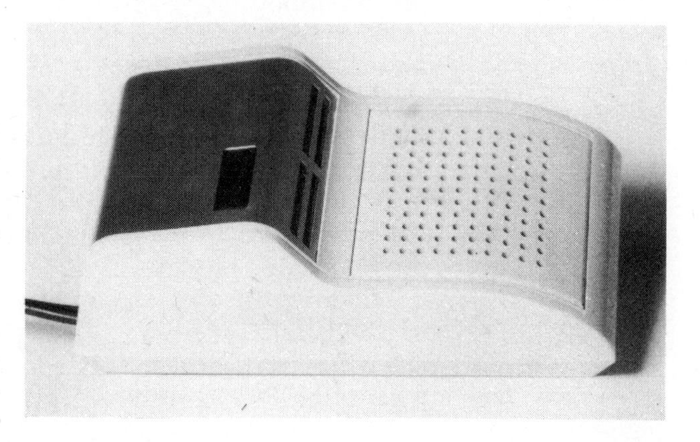

Fig. 3/3 A gas leak alarm which may work from a 12 volt battery or a 240 volt supply will detect and warn of the presence of flammable gases, including carbon monoxide, long before they reach a flammable or explosive level; a safety must if you use L.P. Gas equipment. A simple aerosol-spray type of detector is also available which could be useful for spot checks in the home workshop.

modern and convenient source is to use liquified gas in portable cylinders. These may be the small "Camping Gaz Lumogaz C200" lanterns, giving a light equivalent to about 80 watts operating from a disposable cartridge. The same type of cartridge source can also be used for heating a small workshop using the Gaz "Super Baby C200" portable heater rated at 1000 watts. Other burners are suitable for brazing, etc.

Where the workshop has more regular use, larger capacity cylinders may be found an advantage. A study of "Calor" and "Sievert" catalogues will show the appliances available in each of these ranges.

It may not be generally known amongst those readers who live in towns, that the various national gas boards often have facilities for supplying cylinders of liquified gas to domestic users who are not in a "piped" gas area. If you intend making fairly regular use of your workshop it might be worth approaching your local gas authority to see if they will supply them for your workshop.

In using this type of equipment, safety is of paramount importance. It is well worth while making supports and stands so that lamps and heaters run absolutely no risk of toppling over. When the workshop session nears its end the heater should be turned off, both to save expensive gas, but also to ensure that the burner is cool before you vacate and close up the workshop.

Some readers may not be very sure of just what the "gas" cylinders they purchase contain. Liquified petroleum gas or LPG for short is a fuel obtained from petroleum deposits or during the manufacture of petrol. It has the great advantage that, under pressure, it becomes a liquid and may thus be stored, taking up only about a two-hundredth of the space it would occupy as a gas.

When the tap at the top of the container is opened, the reduction of pressure causes the fuel to issue as a gas which has a very high calorific value (i.e. contains a lot of heat).

Fig. 3/4 Various forms of "cartridge" containers are available for gas equipment. These Convertor Units are convenient in that they convert any of the pierceable butane cartridges into self-sealing units making it possible to transfer them between various appliances.

When handling any liquid or gaseous fuel common sense and care are necessary. However there is no danger as long as two important facts are borne in mind. These are, first, that heating the cylinder containing the LPG will cause the pressure therein to rise. Under normal circumstances the cylinders are amply strong enough to withstand likely increases in pressure but if, for example, a flame were allowed to play on the cylinder the pressure might rise to a dangerous extent.

Second, the gas issuing from the container is HEAVIER than air, hence, should it not be consumed and escape, it will sink to the floor or other lowest point, for example it will run down the stairs! Invisibly of course! So make sure that all the areas where the gas is used and stored are well ventilated.

Various types of "gas" detectors are available which give a warning long before the gas reaches a dangerous level of concentration. These are usually mains or battery operated. However, for the simple workshop using such appliances, a cheap and simple aerosol spray detector is available.

Paraffin

In the days before the advent of the liquified gas container, paraffin, or kerosene to our friends in the USA, would have been our choice for light, heat and perhaps even for power in driving a small, horizontal, single cylinder engine, numerous versions of which were advertised in early issues of the MODEL ENGINEER.

The so-called hurricane lantern is still for sale, and useful where no other sources of illumination are available; while the pressure type or "Tilley" lantern may still be available from agricultural implement suppliers.

Fig. 3/5 Electricity for light, heat and power is very convenient and this blower heater – widely used for greenhouses – is excellent for the home workshop.

Heating the outside workshop, using an up-to-date domestic paraffin heater of the convector type is convenient. But there must be efficient ventilation, because the combustion of paraffin will release water vapour into the air which must be able to escape to atmosphere or it will condense on the tools and machinery with the risk of causing rust, especially if the workshop is infrequently used.

If you do have to rely on paraffin as a source of heat for your workshop, you should take notice of new regulations covering the design of heaters. The new designs are covered by a law which states that such an appliance should not go out if it is standing in a draught, but on the other hand will not go on burning if it is accidently tipped over.

The information to look for on the heater is the statement that it conforms to British Standard (BS) 3300, or alternatively carries the Oil Appliance Manufacturers' Association symbol.

In addition noxious fumes should not be given off by such heaters, but you must be using the correct kind of paraffin. So when buying fuel look for the paraffin symbol (a double row of flames) and the British Standard 2869C1.

Low voltage electric lighting

We may take a leaf from the book of our caravanning friends and make use of low voltage lighting supplied from car batteries. Strip lighting is available for both 12 and 24 volt supply. The fittings are neat and efficient. The interior of our local rural mobile library van is illuminated by fluorescent tubes of 2 feet long and 80 watt output supplied from a pair of car batteries coupled in series.

It is not always the quantity of illumination that matters in a workshop, but having the light in just the right position. Bench and machine lighting can be contrived using automobile bulbs and fittings and have the added advantage of complete safety. It may be of interest to readers without industrial workshop experience to know that the individual lighting fitted to machine tools has, by law, to be of the low-voltage type. Thus by using this form of lighting in our garden workshop we are conforming with correct safety requirements. If you then set up a car generator to re-charge the batteries, especially if it is driven by a lovely steam engine and boiler, you will not only have a workshop but your own power house as well!

If you have an interest in energy saving you might even have a wind generator.

These ideas are in no way ridiculous, especially if the workshop is remote from the house. I well remember the very first model engineer's workshop I ever visited as a young lad. The machines, lathe, toolgrinder, drill, etc., were treadle operated, but there was also a lineshaft across the centre of the shop driven by a horizontal steam engine with coal or wood fired boiler. Thus when a long evening session was envisaged, the boiler was lit up, keeping the shop really cosy. Work progressed with occasional interruptions to fire the boiler and check water level. All this fun in a home-built wooden shed 8 feet by 6 feet.

Finally, these latter comments raise an important point which is that even in this day and age there is no reason at all why any machinery we have in the workshop, small lathe, drilling machine, tool grinder, even circular saw, should not be treadle operated. If we are reasonably healthy the exercise will certainly warm us and do us a lot of good, especially if we have a sedentary occupation – and think what you will save on the fuel bill!

4. BENCHES, CUPBOARDS & SHELVES

If I were starting a workshop from scratch all over again, the first thing I would do would be to build a really substantial bench. Even if initially it has to stand outside, protected from inclement weather by a sheet of polythene, the possession of a stout bench fitted with the usual necessary accessories is a tremendous asset.

Benches

Of course, like everything else connected with workshops, the size and style of the bench must be planned. Make it of a reasonable size and if it is intended to be subsequently built into the structure of the workshop, plan for that contingency at this stage.

The overall size, i.e. length and depth, will depend on material and space available, but while 1½ to 2 feet is sufficient depth, try to make the length as great as possible. Height depends on whether you are predominantly interested in working in wood or metal. The most useful method if, like me, you have only one bench, is to make it of a suitable height to suit your stature for planing wood, then make a raised portion to take an engineer's vice with the top of the jaws up at elbow level for filing. This can be at one end of the bench, leaving the majority of the length for planing long planks of wood. Woodworking is very greedy of space, metalworking is not!

It would serve no useful purpose to propose and

Fig. 4/1 Readers who did woodwork in school will remember the traditional type of double-sided bench that was used. This has all the requirements that make a good work-place. It has ample top area and is strongly built with plenty of weight to give stability.

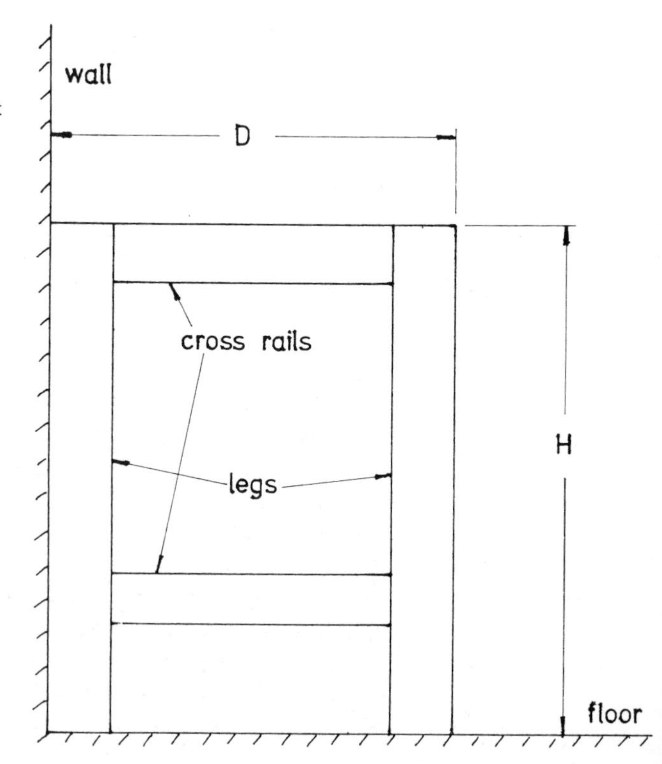

Fig. 4/2 Planning the bench.

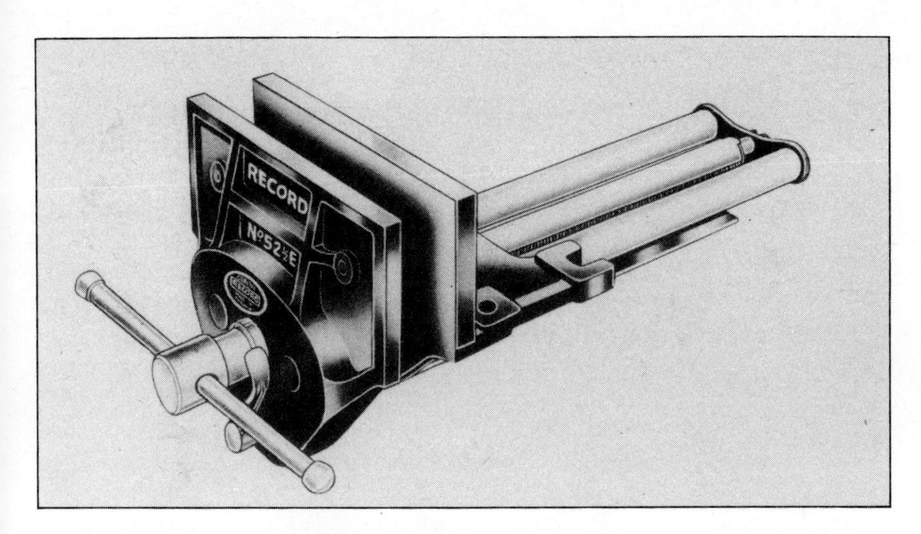

Fig. 4/3 Woodwork vice.

describe in detail a design for a workbench, because such a workshop feature is a very individual and personal thing, that depends on many variables. These comprise, the space available; the situation, i.e. can the bench be built-in or is it to be free-standing; the type of work to be carried out, i.e. "light" or "heavy"; the available material; and last but probably most important the tools and skill of the builder.

Regardless of the type and size of bench, the chances are that we shall require some form of legs and leg framing, as illustrated in Fig 4.2. The height "H" will depend on our physical height and whether woodworking or metalworking is going to predominate. If, as seems likely, it will be a bit of both, then make the bench suitable for woodworking and pack the mechanic's vice up on wooden blocks to raise it to a suitable height for filing.

To assess accurately the height "H" to suit YOU, equip an assistant with a tape measure, then, while you hold a woodworking plane in a suitable position for planing, get him to measure the distance from the sole of the plane to the floor. This will give you a custom-built bench at which the planing of wood should be comfortable, bearing in mind that a woodwork vice (Fig 4.3.) is fitted IN rather than ON the bench, see Fig 4.4.

If only light metalworking is envisaged, the mechanic's vice can be bolted to a stout timber sub-assembly which is held in the woodwork vice. However sooner or later you will want a properly fitted engineer's vice and this should be permanently fitted, blocked up so that the top of the vice jaw is on a level with your elbow. You will find this just about right for effortless filing and hacksawing.

The time you spend in the workshop is supposed to improve your well-being both mentally and physically. This won't be the case if your bench is uncomfortable to work at, so don't be satisfied until you get it right.

The bench depth, that is the distance from back to front, depends very much on available space, 450 to 600 mm (18 to 24 inches) is about right. Anything wider and it tends to be used as a repository for tools, narrower and it's like working on a shelf.

Referring back to Fig 4.2, the general construction of any bench will consist of two or more such frames, joined together by longitudinal members. Hence the making of these frames would seem to be the most important item.

As you are likely to possess a few simple woodworking tools, let us consider bench building in this material.

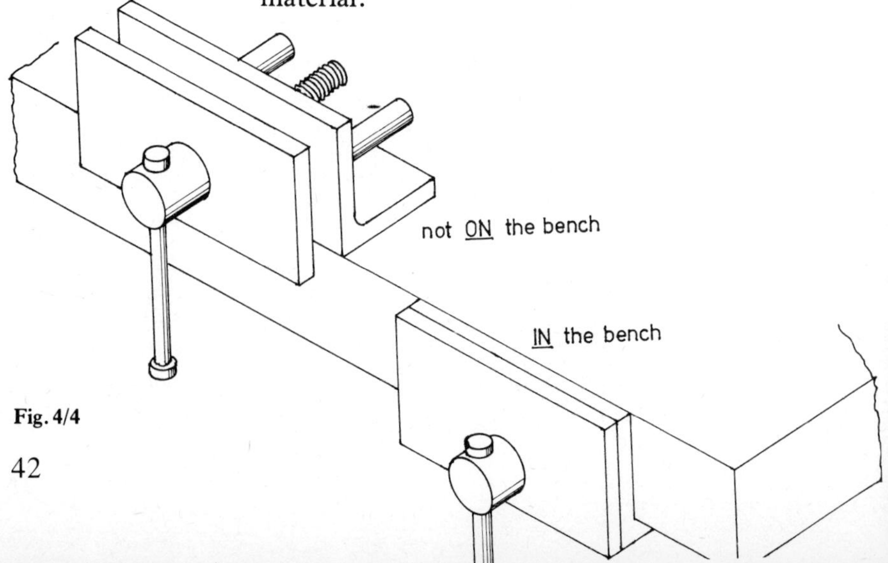

Fig. 4/4

The simplest way to make up these frames, using plank or board type timber, and no more tools than a handsaw and a small drill, is illustrated in Fig 4.5. The timber is simply sawn to length; two legs and two cross-members being needed for each frame. The structure is drilled as shown and bolted together using standard coach-bolts (Fig 4.6), obtainable from any ironmonger or DIY establishment. The bolts must be long enough to go through the two thicknesses of wood plus sufficient to take a washer and nut. Bolts of $\frac{1}{4}$ to $^{3}/_{8}$ inch diameter should be suitable and two at each joint is sufficient. Some glue or old thick paint applied between the timber before final assembly will help to make a firm joint.

If the timber available is of heavier section, something like old door frames, joists or rafters, we can still bolt the assembly together but the technique is slightly different, as shown in Fig 4.7. Here the cross-members are butted against the legs and located by two wooden pegs (dowels). The joint, which may or may not be glued, as desired, is pulled up tight by a bolt, washers, and nut, fitted as shown. The normal coach-bolt is not suitable for this application because a hexagonal head is needed on the outside so that we may apply a spanner to tighten the joint.

In the past, I have used screwed rod with a couple of nuts and washers for this type of joint in various timber structures. Screwed rod, or "studding", may be obtained in various thread sizes from a local engineering stockist or by mail order from Whiston, New Mills, Stockport. It is a useful mechanic's workshop material as, by its use, screws of any length can be made up either as studs or as set screws, by brazing a nut on one end to form a head.

Alternatively we can produce a neat frame if the joints are made by "halving" the timbers into each other. This is shown in Fig 4.8.

Again coach-bolts and glue or old paint are used to effect the joint. Mark them out carefully and try to ensure that they are a firm, even tight fit. When cutting this type of joint, take care that you keep all parts of your anatomy behind the cutting edge of the tool!

If you have a mortise chisel of a suitable size, you may like to do the job in the traditionally correct manner, that is by using mortise and tenon joints. The details for this are shown in Fig 4.9.

Normally, the width of the tenon is made about one third of the thickness of the timber, but if in doubt, or if the leg members are thicker than the cross-rails, the tenon may be made appreciably wider.

Mortise and tenon joints are glued together, but a considerable increase in strength may be obtained by wedging and/or pegging the joints. All these details

Fig. 4/5 Simple bench frame.

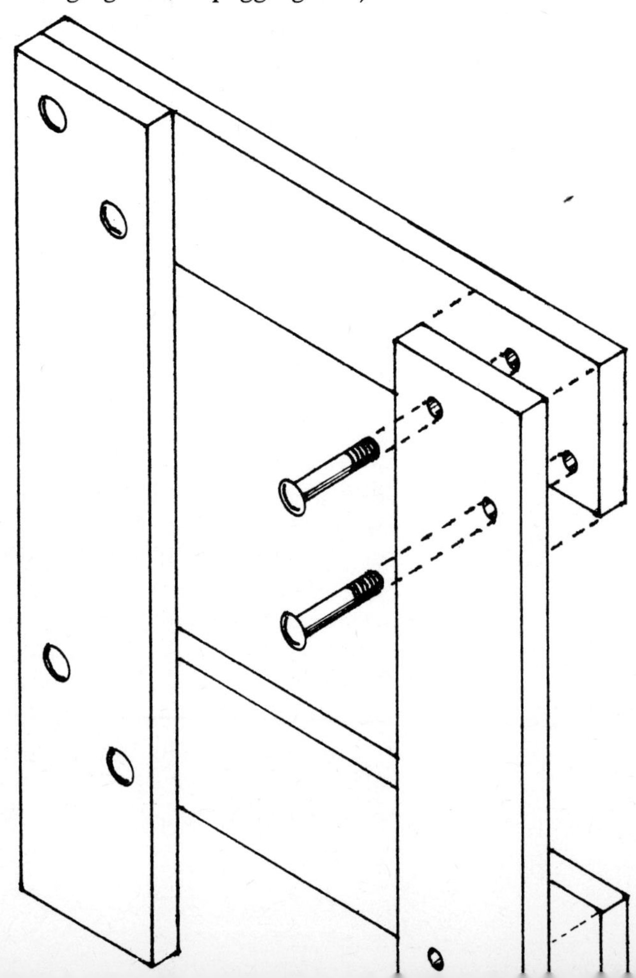

Fig. 4/6 Coach bolts and screws.

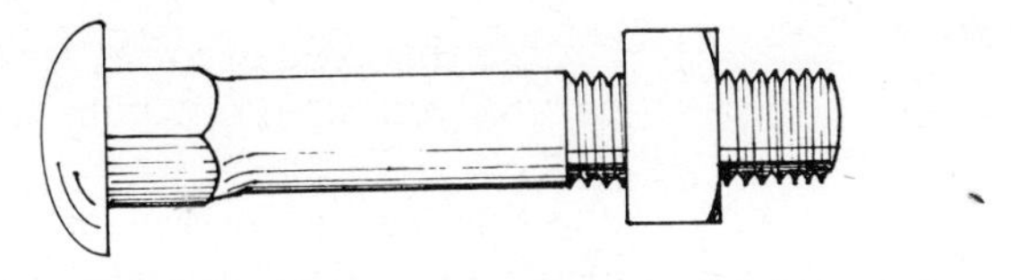

Coach Bolt sometimes referred to as Cup-Square-Square, ie Cup-head, Square-neck, Square-nut.

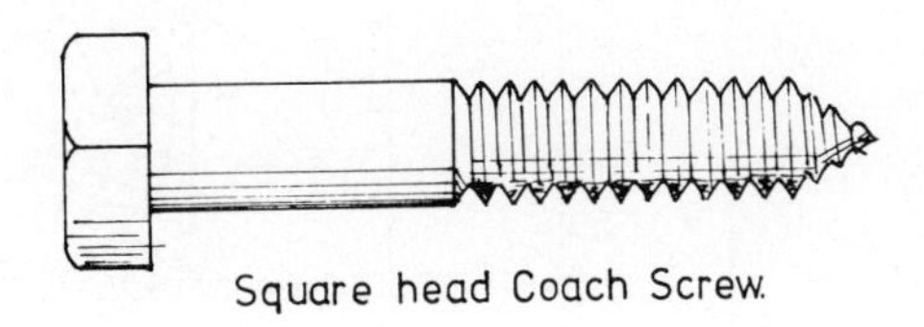

Square head Coach Screw.

are glued as well, thus binding the whole joint firmly together.

The longitudinal members that tie these individual leg frames together, may be fitted in a similar manner to that employed for the cross-members. Or we can save on timber by using the top and an under-shelf to separate the leg-frames. If we do the latter, it is worth while fitting a diagonal strut across the back to maintain stability.

Use secondhand material

Whatever the bench top material may be, it pays to cover it with a sheet of hardboard, fitted by small nails called panel pins. This makes an excellent surface which, when it gets worse for wear, can be removed and easily and cheaply replaced.

All the above presupposes that you have had to start from scratch when building your bench. Before you do, have a good look round to see if there is perhaps some item or old piece of furniture that might be modified to form the basis. Even if the furniture is damaged, simple repairs are possible by gluing or the use of screw plates of various forms. Worms can be dealt with by treating with one of the proprietary solutions. A corner of my own workshop

Fig. 4/7

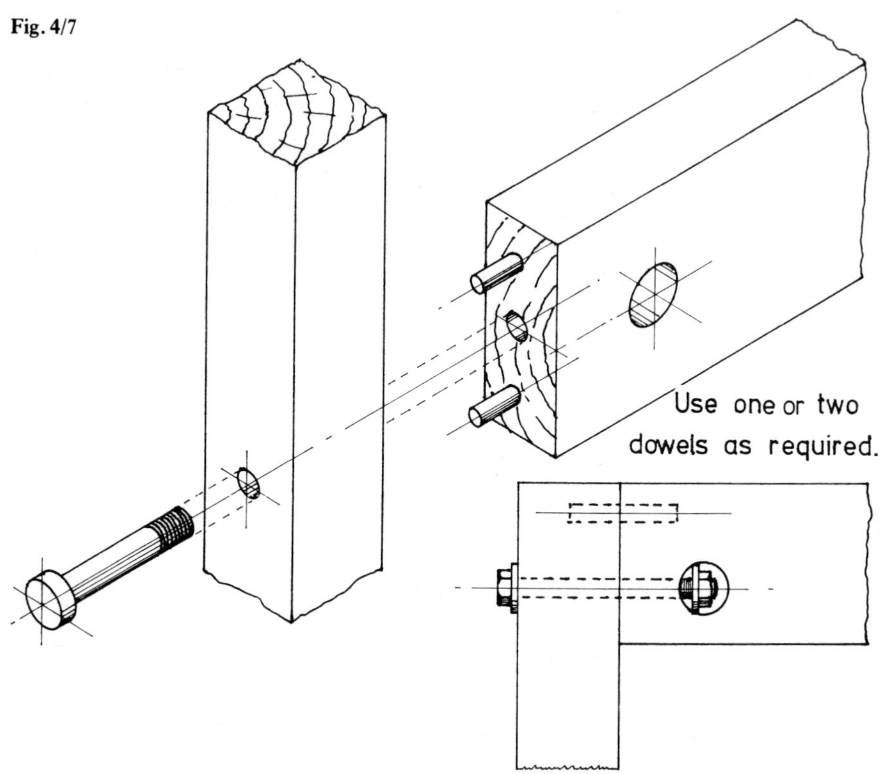

has the bench built up on an old pine chest, bought a quarter of a century ago for 10/- (50p). Unfortunately my wife now appreciates its value in this "stripped pine era", and the excuse that my bench will fall down if I remove it isn't going to be accepted much longer – is nothing sacred?

With material the price that it is, you will no doubt have to get what you can to make the bench, but stout second-hand door frames can often be obtained from a local jobbing builder and are ideal material for the bench framework. From the same source one might also obtain an old door that could be turned into a bench top. Even the old-fashioned, panelled type, no longer appreciated in modern houses, will make, when covered with chipboard, etc., a stout bench top.

Alternatively, a visit to a scrap metal merchant

may bring to light an old cast-iron or welded steel frame which with ingenuity can form the basis of a strong bench that will last a lifetime and beyond. In the past I have had benches made of everything from old air-raid table shelters to off-cuts of larch fencing posts. The latter gave the workshop a delightful "piney" smell.

If the workshop has a concrete floor, make the bench of such a height that a duckboard can be accommodated for standing on. This will keep your feet much warmer than strips of carpet. A duckboard is simplicity itself to make as the sketch shows.

More than just a workbench

If you don't want to take time to make a bench, they are available commercially; beautifully made to various designs and with a range of fittings that make them ideal for a wide range of craft activities. If you are setting up a home workshop for a specific purpose you cannot do better than choose the correct bench from the wide range offered.

For many people the Scandinavian style of workbenches will pose an unfamiliar sight. Almost anyone who has used one however will agree that it

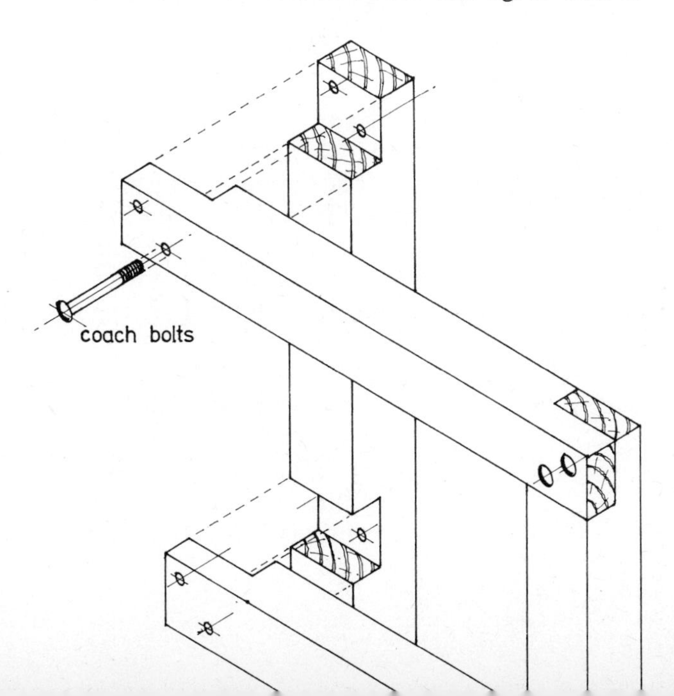

Fig. 4/8 Frame made using halving joints.

Fig. 4/9

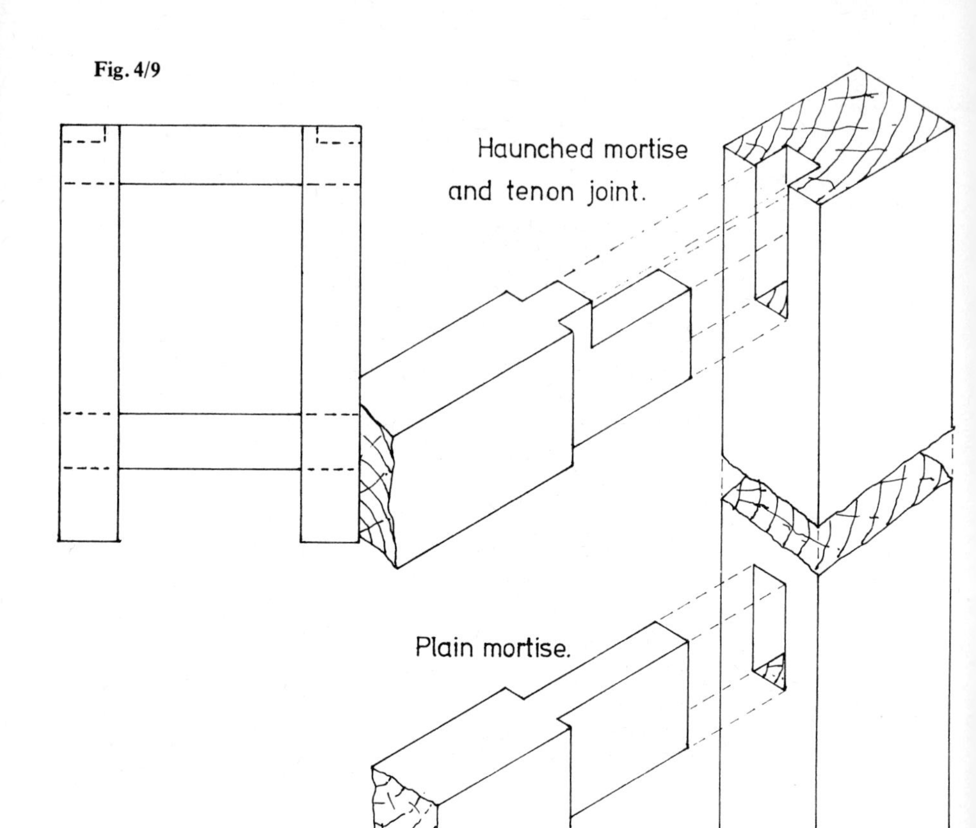

affords excellent facilities. The Lervad benches are a good example.

The first unusual feature is the long narrow bench top. This is designed to provide easy access to the workpiece from three sides and so obviates the constant repositioning of work so often necessary on the more conventional rectangular shaped bench. These are benches to be worked *round*, from virtually whichever angle the user finds most convenient.

Another outstanding feature is the number of different ways work may be held. Many Lervad benches have no fewer than three different clamping facilities which, used in various combinations, ensure that any piece of work, whatever its shape or

Fig. 4/10 (a & b) **Always remove nails from secondhand timber as it is dismantled. If time does not allow this, bend them over as a temporary safety precaution. Photo: Stanley Tools Ltd.**

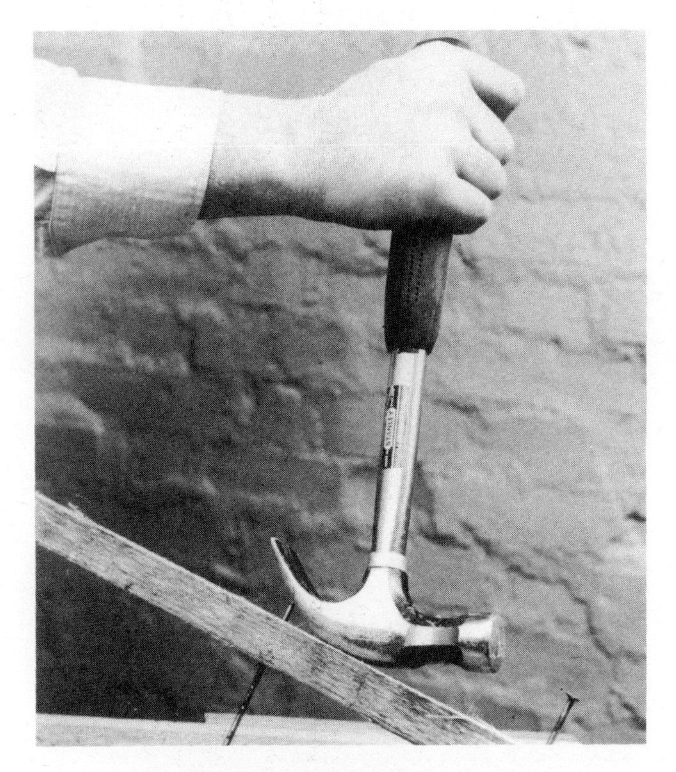

Fig. 4/11 Cutting mortise holes. Photo: Stanley Tools Ltd.

Fig. 4/12 For safety's sake be careful where you put the G-clamp. Photo: Stanley Tools Ltd.

size, may be held safely and perfectly firm.

Benches may have full width tail vice, integral tail vice, shoulder vice or either a Record or Jeros pattern vice. Almost all Lervad benches have in addition a bench dog clamping system, which extremely versatile facility greatly improves the efficiency of the benches.

Benches equipped with the full width tail vice have a double row of bench dogs providing a four-point hold for workpieces.

The work surface itself is also of note. Principal timber used in making Lervad benches is prime Danish beech, which is used for the bench tops. This is seasoned, kiln dried, allowed to normalise and then totally immersed in linseed oil before being lacquered. This care serves to ensure the perfect flatness of the working surface under all normal conditions.

In addition to benches for woodworking, Lervad also manufacture benches specifically for metalwork, tables for craftwork, a range of fine weaving looms and storage cabinets. The unfailing high

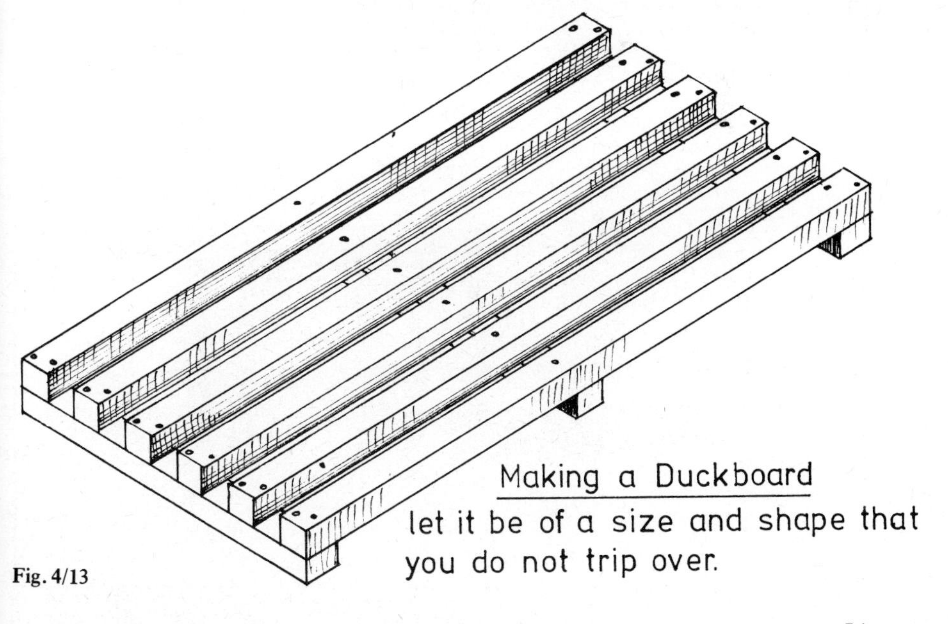

Fig. 4/13

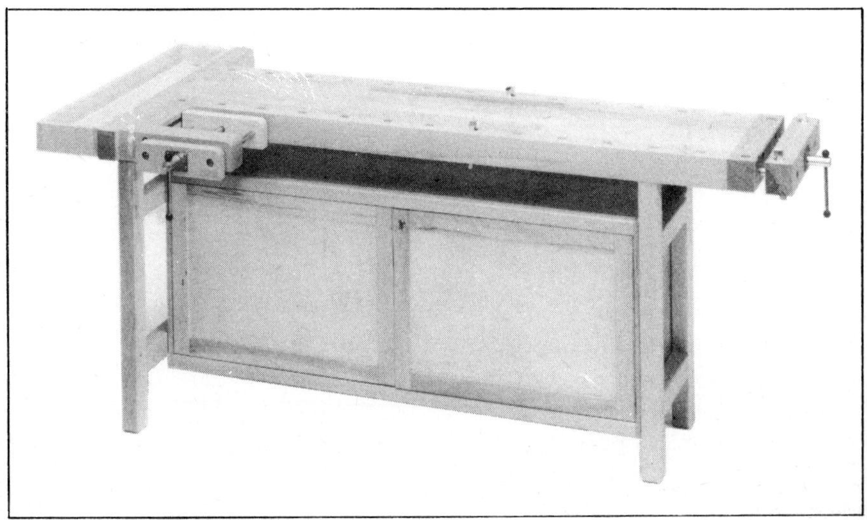

Fig. 4/14 For those who prefer a conventional work surface, the Lervad 605 bench provides plenty of room and multiple clamping devices.

quality of design and finish on this equipment has led to its adoption in many schools and colleges, and indeed for any keen craftsman it represents a fine investment.

Cupboard accommodation under the bench is an asset. The 611 Lervad comes so equipped and though under a narrow workbench top it can be shelved to hold a surprising number of tools.

Fig. 4/16 The Lervad 610 bench, beautifully constructed of top-grade Danish beech, with a worktop kiln dried and oil sealed to obviate shrinkage and warping. The bench is of typical Danish design and boasts a combination of three different clamping devices, including a double row bench dog system.

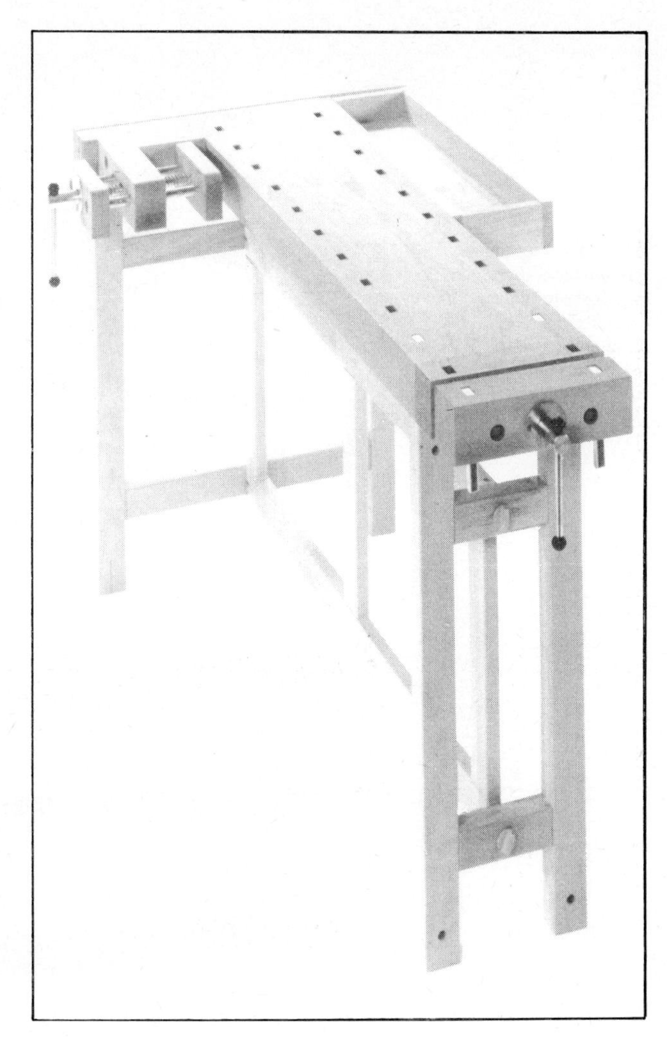

Fig. 4/15 Bench No 611 from the Danish company, Lervad, is supplied with cupboard accommodation to provide the perfect workplace for any keen woodworker.

Fig. 4/17 **Brackets may be made to any reasonable size, keeping to the proportions given.**

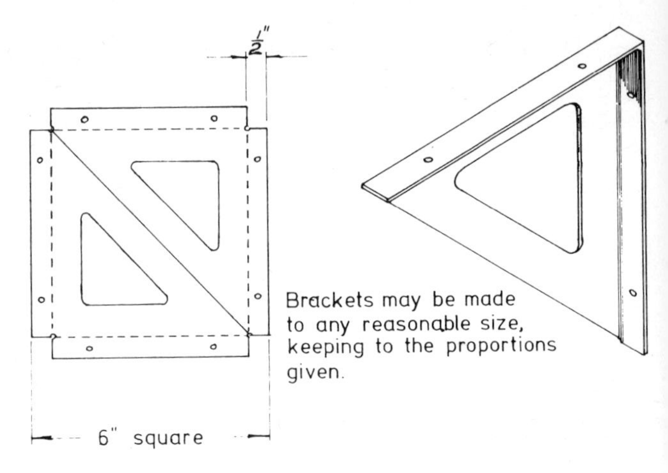

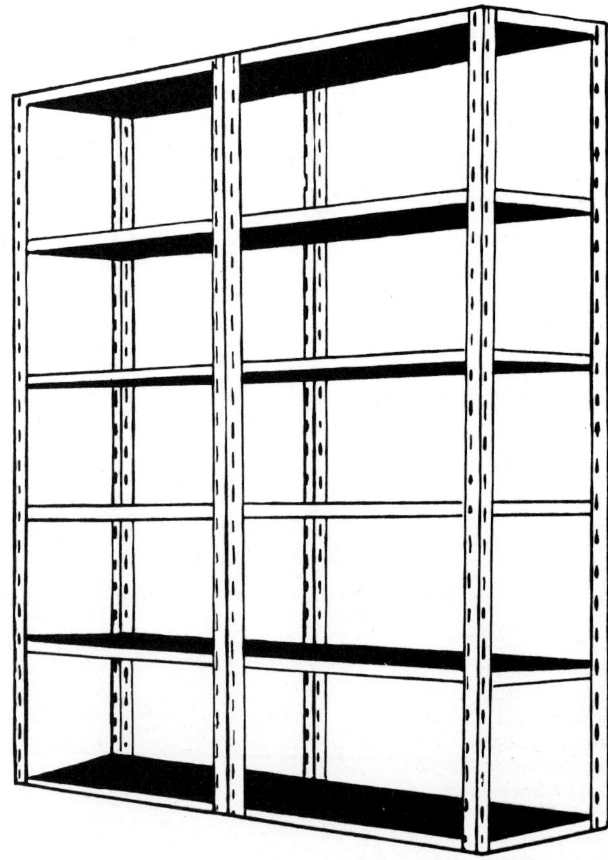

Fig. 4/18 **This plastic coated steel shelving is simple to erect and provides instant availability to the workshopper in a hurry. It is widely advertised.**

Cupboards and shelves

The large empty space within a bench will immediately recommend itself for tool, material and work storage. If the necessary panelling and shelving is substantial and firmly fixed, this will considerably strengthen the bench, making it good and weighty so that it doesn't walk away when you work at it. But don't take the cupboard to the floor; leave "toe-room" under the bench or you will find it uncomfortable when working.

While thinking of shelves, it is my practice to make narrow shelves of solid timber and wide shelves, slatted. I find that the latter then do not collect dust and dead spiders quite so quickly and if the shelf is above eye level it makes the contents more readily seen. Plus the fact, of course, that there is a saving in timber. Simple and easily made shelf brackets are

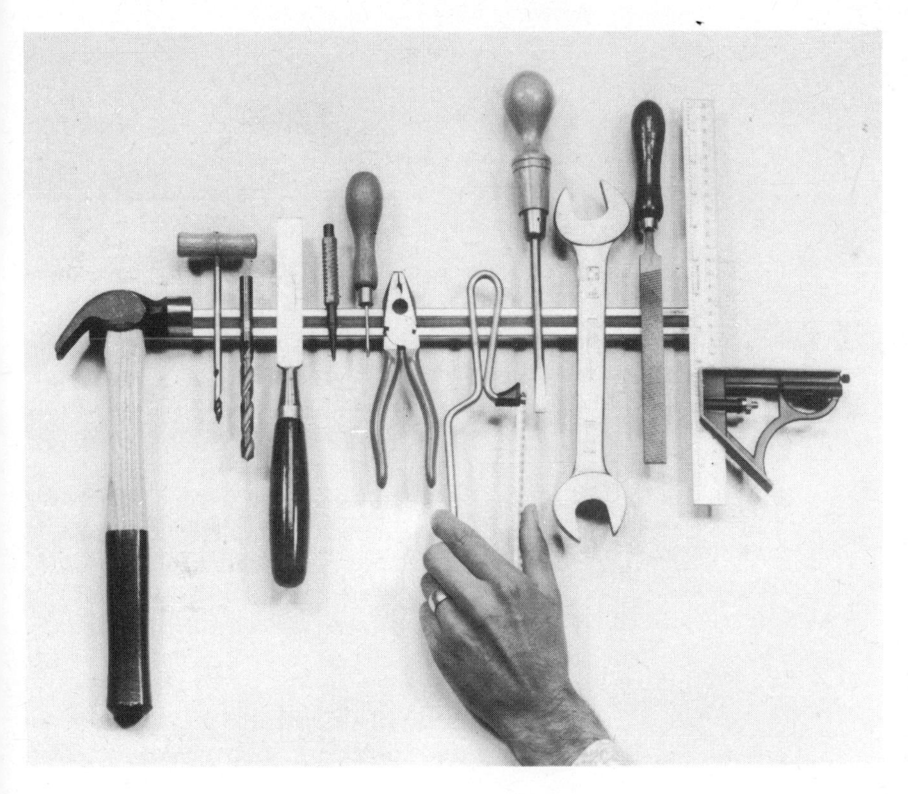

Fig. 4/19 For tools that require to be instantly at hand the POWERLINE Magnetic Tool Rack is easily fitted to most surfaces. It is in 12 and 18 inch lengths but may be built up to form a rack of any length.

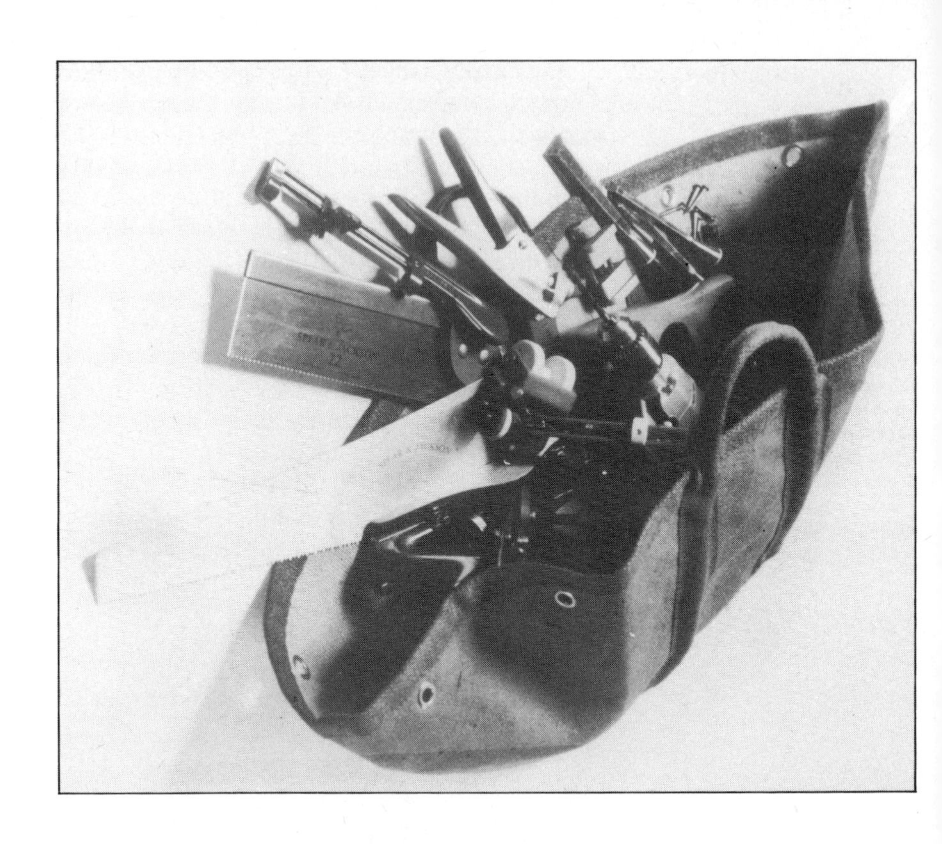

Fig. 4/20 Arrangements for storing tools can range from the traditional carpenter's tool bag to plastic tool-boxes with itemised compartments.

shown. They can be quickly made from any sheet metal and are quite sturdy. A few hours' work with tin-snips, drill and file will produce half-a-dozen pairs at minimal cost.

Free-standing shelving is often desirable, and in fact, in a flimsy workshop, may be an absolute necessity. Very strong plastic coated steel storage equipment of this type is available at a very reasonable price, with the advantage that it may be easily taken to pieces and reassembled elsewhere should the need arise. In fact, I suggest filling up the shelves as quickly as possible before your wife decides that the stack of shelves would be more convenient in her kitchen!

The sloping walls of attic rooms often cause problems when trying to put up shelves. An idea suggested by the makers of "Tebrax" shelving uses their shelf supports fixed to a front frame. In a workshop the front would probably be left open, but the curtain idea is useful if the attic room may occasionally have to double as a bedroom.

Sets of drawers and cupboards no longer good enough for domestic purposes can have a new lease of life in the workshop but, unless there is a very special reason for not doing so, I suggest that you remove the doors. They are not really necessary and in the close confines of a small workshop can be a confounded nuisance. There are exceptions however such as if

Fig. 4/21 Somewhere to store your portable electric drill and its accessories, and, at the same time, easily transportable for these jobs outside the workshop. The DRILL-MATE is strongly made and carries any ¼ or 3/8 inch drill.

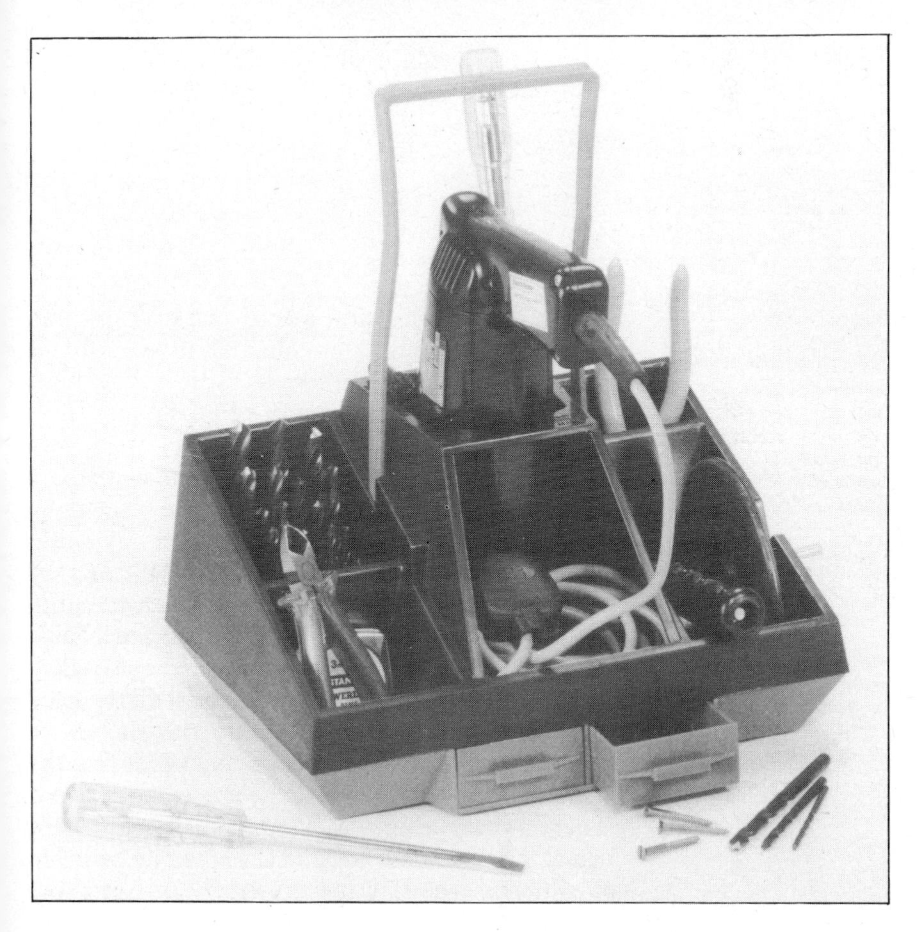

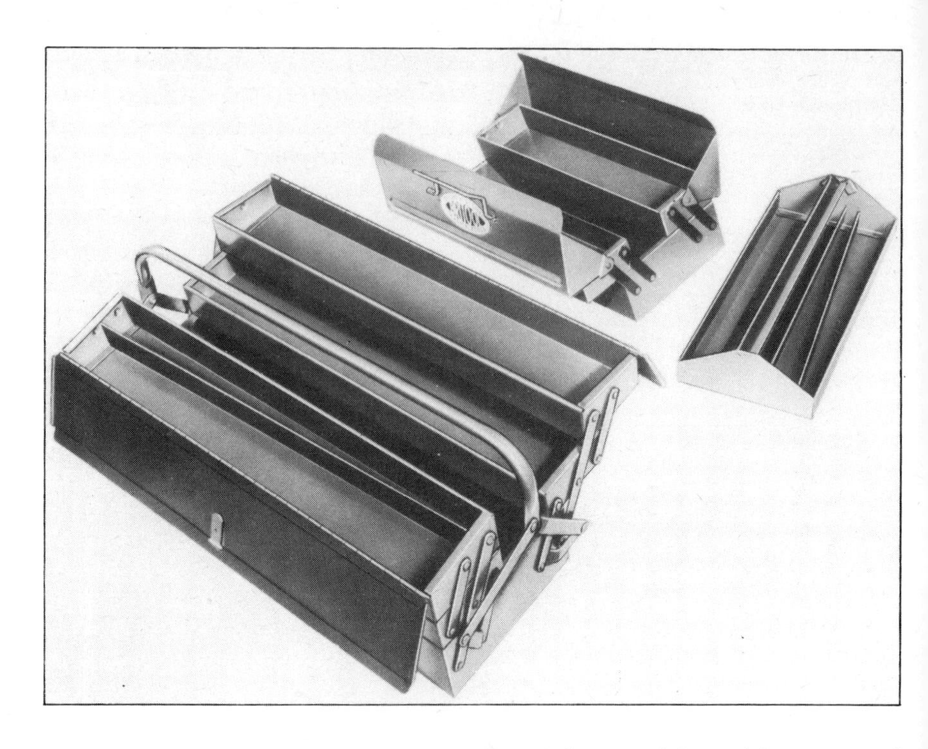

Fig. 4/22 The engineer's tool-box, which can come in many variations, some lockable, all portable and always handy for carrying tools to the job in hand.

you have a cupboard for special marking out and measuring equipment and you wish to keep it free from risk of dampness. In which case leave the doors on and line the shelves with VPI (vapour phase inhibitor) paper.

If dampness is a real problem in your workshop, the only guaranteed answer is to use the place more often! Bright steel will eventually rust in any workshop if it is infrequently used. However, if you must leave it unused for some time, lightly smear bright surfaces with a water repellant oil such as Shell Ensis 252 or similar.

Built-in fittings

Earlier we mentioned the possibility of building the bench (and other fittings) into the workshop. This, if you have a cheap, flimsy wooden shed, is an excellent scheme. This is in no way a criticism of such sheds. You get what you pay for and with some of those advertised it would be difficult to buy the new timber for the price quoted for the finished shed.

By building in bench, shelves, cupboards, etc., as part of the shed framework and firmly screwing everything together, you not only save money but also considerably stiffen up the complete structure. One point always to remember is to treat every bit of timber with a preservative against fungoid decay and woodworm. If you don't mind the dark colour and the smell, creosote based preservatives are cheap and easy to apply. Alternatively there are types which leave the timber in its natural state and odour free.

Portability

The opposite of built-in storage facilities is needed when the workshop (and you) is called upon to demonstrate its prowess by carrying out jobs of maintenance and repair about the house. Then the running back and forward between work and workshop can be both frustrating and time-consuming.

The normal type of toolbox usually results in having everything tipped out on the floor to find a particular small tool, but a portable tool storage transporter made of heavy duty styrene has every tool visible plus ample drawer space for small nails, screws, etc.

A similar piece of equipment, Fig 4.21, for $\frac{1}{4}$ and $^{3}/_{8}$ inch electric drills together with the necessary accessories is valuable for workshop storage and portable use of these popular tools.

5. VICES & WORK HOLDING EQUIPMENT

Without its various accessories a bench is limited in its use and possibilities. The first fitting is obviously a substantial vice and whether this is of the woodwork type or for engineering, depends on the branch of practical work in which you are primarily interested. In my own case, although my workshop may be described as being predominantly for light engineering and my bench is fitted with a 4½ inch mechanic's vice, I also have a 9 inch woodwork vice opening to over 12 inches. This I find invaluable for holding large diameter pipe when boiler making etc., as well as its more normal purpose for woodwork.

If you are only going to do occasional woodwork, a suitable vice is not difficult to make from simple materials, to the proportions shown in the sketches. Although the photograph shows the vices made from castings, for a one-off, pieces of steel plate form an ideal alternative. I have made such vices up to a 12 inch jaw size.

Vices for woodworking are fitted with the jaws just below the level of the bench top and have the iron jaws faced with wood to protect the work. Use coach bolts to fit the vice, with the bolt heads sunk below the level of the bench top. Plug the resulting hole with wood if you wish.

The engineer's vice is mounted on the bench with the jaws up at elbow height. The vice jaws are serrated to help hold work firmly, but in doing so unfortunately

Fig. 5/1 A group of student-made vices using simple castings.

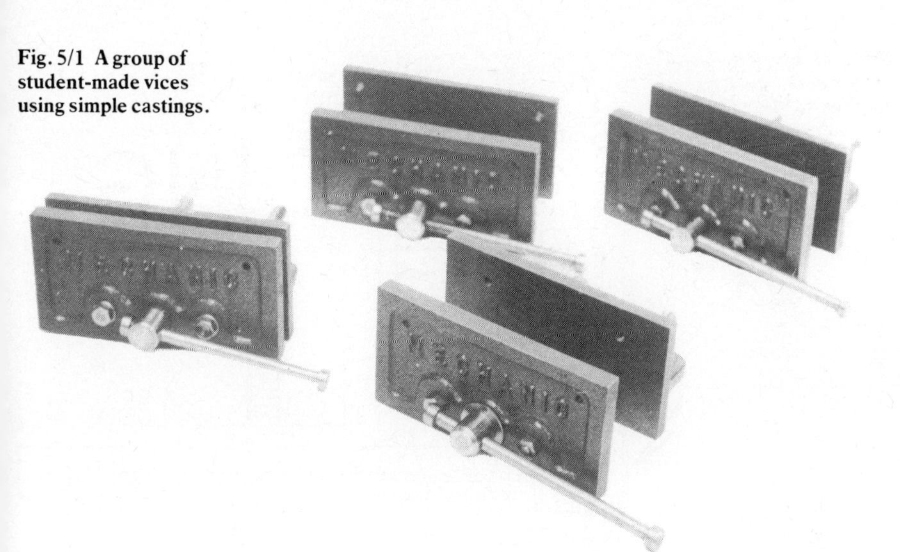

Fig. 5/2

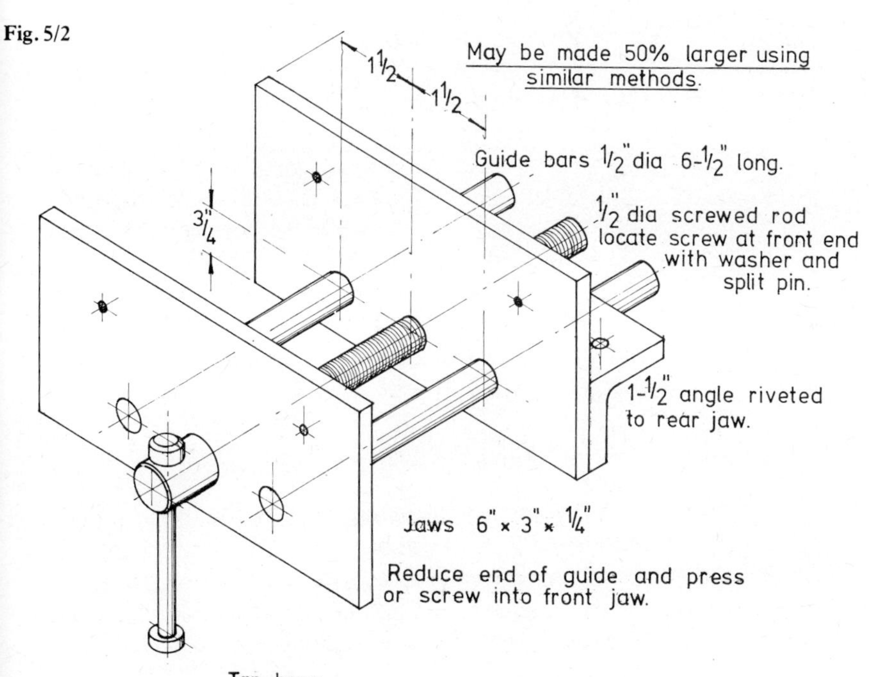

Fig. 5/3 Fixing a woodwork vice. Softwood "faces" to protect your work.

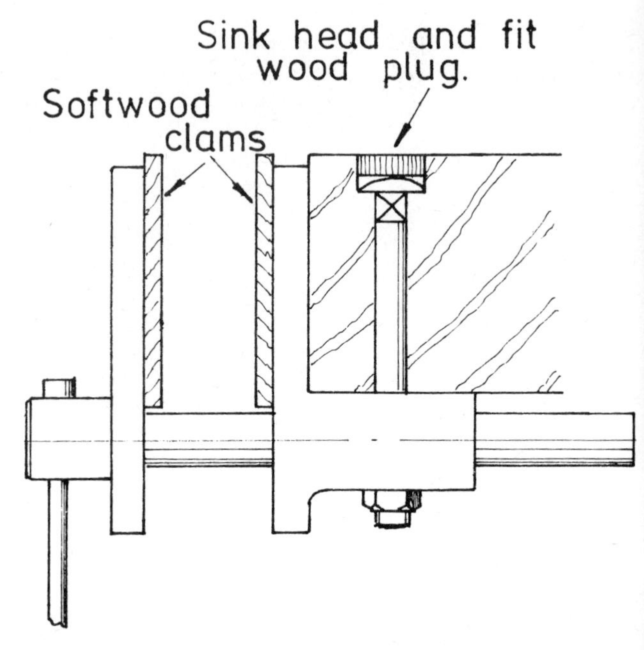

Fig. 5/4 We have mentioned the advantages of attending classes at the local technical college; here a group of students survey the engineer's vices they have made at such a class.

Fig. 5/5 **Commercial bench stop.**

mark the metal surface. For jobs where these marks are unacceptable we use vice clams, bent up from scrap pieces of soft metal, usually aluminium sheet.

Again, the engineer's vice is fitted using coach bolts, but this time with the heads below the bench top and the nuts showing above the vice. The details are given in Chapter 7.

Various adaptors may be made as required to make the holding of certain shapes more convenient. This is especially useful in metalworking when it is desired to hold rod, tube and studding (screwed rod). Always make such devices with a lip or pegs so that they rest on the vice jaws. This saves the need to have a third hand!

When planing the surface of a piece of wood a bench stop is a necessity. This can be simply a short strip nailed across the left hand end of the bench top. It will however, not suit any thickness of timber and the nail heads are sure to foul the plane blade. It is preferable to fit a commercial type of stop or make one as shown fitted against the bench leg.

Fig. 5/6

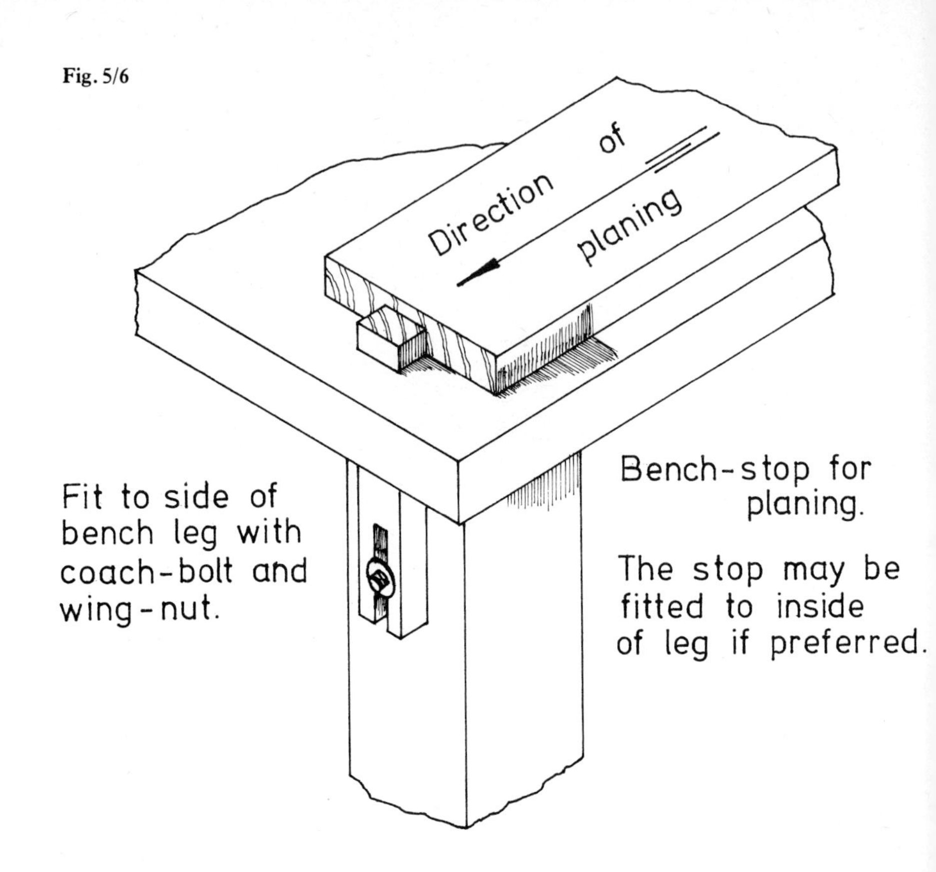

Fig. 5/7 Vices come in many forms but all with the same purpose of holding work so that it may be drilled or shaped.
This Machine Vice is popular for use on drilling machines, shaping and milling machines, etc.

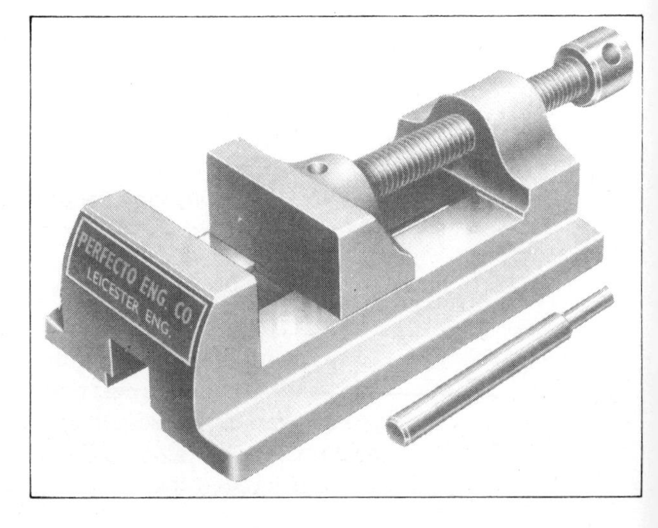

Fig. 5/8 Close-up view of the double bench-dog system on a Lervad bench. When used in conjunction with the other clamping devices on all benches this ensures that all pieces of work – however large or small – can be held securely.

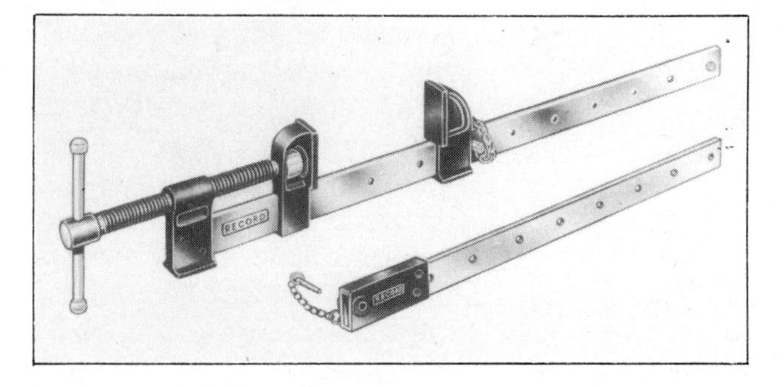

Fig. 5/9 Sash cramp, normally for woodwork, but useful in all types of situations.

Where woodworking predominates, or for portability, the "WORKMATE" adjustable vice/bench is useful, especially the fact that it can be folded up for storage.

Various other forms of clamping devices may be bought or made as the need arises. The most popular and useful of these are described as G-clamps and may be purchased in all sizes. They are not difficult to make and it is worth having a small stock of studding (screwed rod) of a suitable size to suit the range of clamp sizes

Fig. 5/10 Bench holdfast, useful for sheet metal as well as wood.

normally required. If you ever build a boat, you will appreciate the enormous range of permutations possible between capacity and throat-depth.

A type which may be considered easier to make is the so-called "toolmaker's clamp". I say "so-called" because the same type of device, made in wood is used in cabinet-making and for veneering, etc., in which case the jaws are referred to as "chops"! Neither of these is difficult to make in various sizes if commercial screwed-rod is used.

Such simple work-holding and clamping devices are excellent projects to make during a winter session at the evening classes of a local technical college or evening institute. Any newcomer to the skills of owning and running a home workshop is earnestly advised to make full use of the excellent facilities offered by these establishments.

The angle-iron vice or clamp for bending sheet metal illustrated can easily be rigged up by the most unskilled novice and is extremely useful and adaptable. Depending on the size of your bench and the

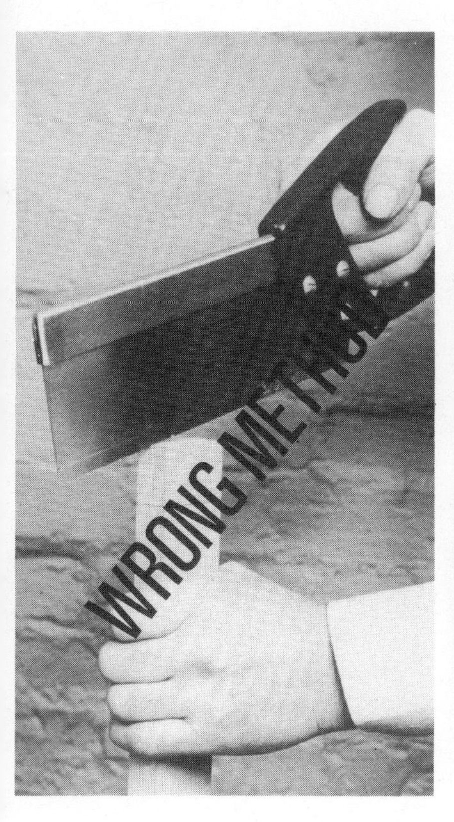

Fig. 5/11 (a & b) **If work must be held high in the vice, never steady it by using your hand; use a cramp to stop vibration. Photo: Stanley Tools Ltd.**

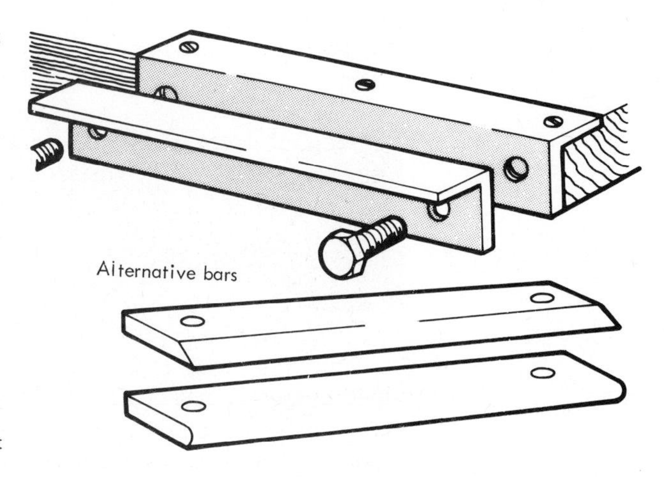

Fig. 5/12 Vice for bending sheet metal; make to any convenient length.

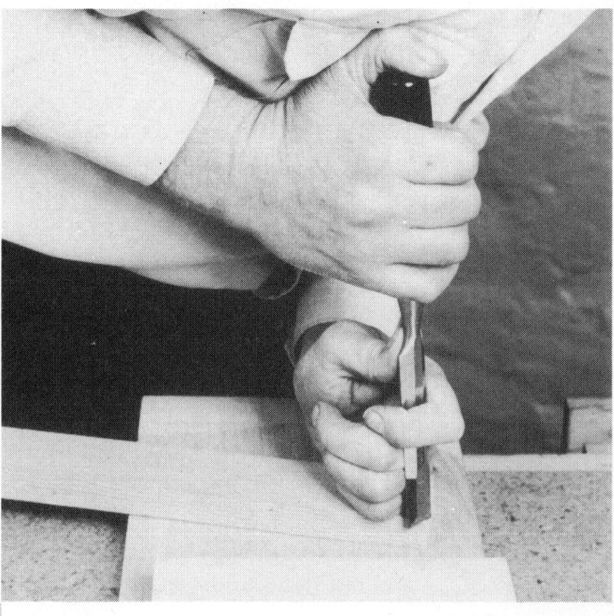

Fig. 5/13 (a & b) A sawing board (sometimes called a bench hook) is easily made to hold wood for cutting with a tenon saw or when paring with a chisel. Notice the correct methods of gripping the tools with all parts of the anatomy behind the cutting edge! Photo: Stanley Tools Ltd.

Fig. 5/14 Sketch of sawing trestles.

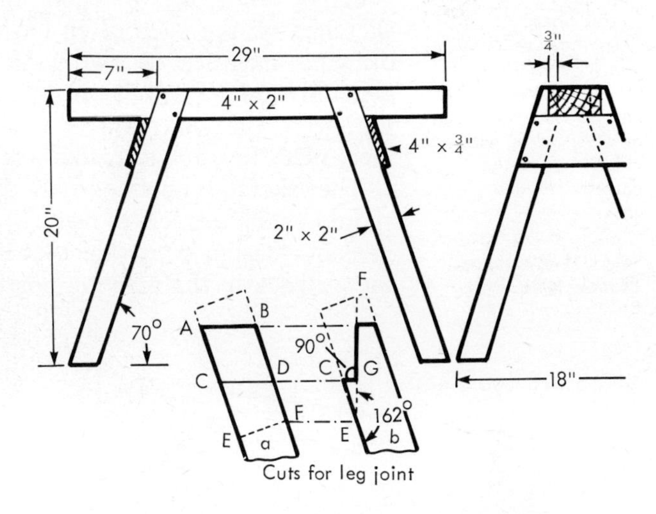

work you expect to do, it can be made from steel angle of any size and of any reasonable length.

It is also useful for holding wide boards of wood for working the ends such as cutting dovetail joints.

Two useful and simply made devices for supporting timber while cutting are the sawing board (or bench hook) and the sawing trestle.

The first is made, preferably in hardwood, by estimating the sizes from the photographs and is assembled with screws or wooden dowels. The trestle, of which a pair is most useful, is slightly more tricky and needs some precise marking out and cutting. Even so a bit of careful adjusting may be subsequently needed for a good fit. The sketch shows those I made many years ago. A simpler type using hinges and short lengths of chain or rope is also illustrated in use, see Fig 6.2.

Workshop Drawing Table

While on the subject of working in wood, the table illustrated has in the past proved invaluable in my own workshop.

That odd drawing which is often found necessary,

generally ends up by being dispensed with, or sketched on the back of an old envelope. Lack of drawing facilities in the workshop or home is generally put down to lack of space, while the bench top, from the viewpoint of cleanliness, is not the ideal place to use a drawing board.

The sketch contains main dimensions which should enable readers to make up a simple and convenient "design department" where you can really say – "back to the drawing board"!

Fig. 5/15 Workshop drawing table.

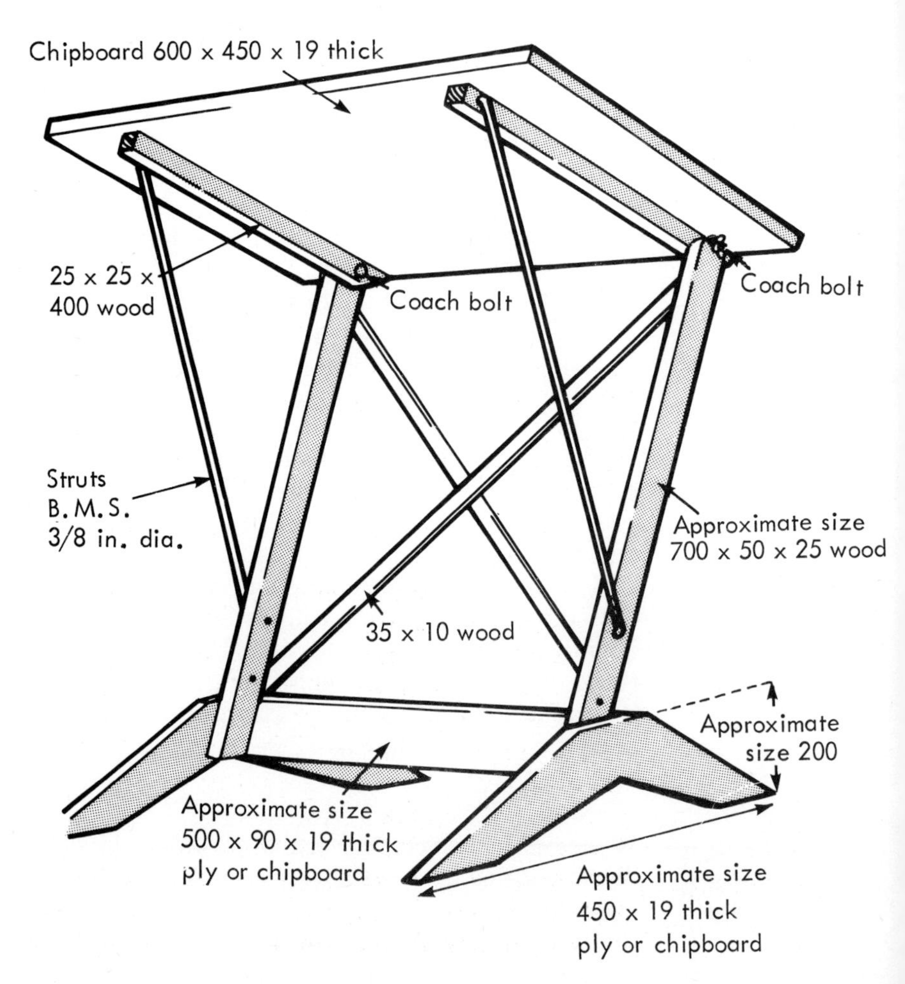

6. BENCH & HAND TOOLS FOR WOODWORKING

Wood and metal-working tools

This subject could well form a publication in itself and so we can only deal with the more important aspects to help you get started. An old illustrated tool catalogue begged from a friendly tool merchant will be useful to show you what is available and its name.

There are few households without some tools, however simple. And no doubt, you are reading this book because you have an interest in tools and like handling and working with them. It goes without saying that it is worth buying the best tools you can afford. This generally means that they should be well made, finished, and "come nicely to hand", i.e. feel right. In addition, cutting tools must be made of the proper grade of steel and correctly heat-treated if they are to work efficiently. This latter requirement is solely in the hands of the manufacturer, so it is important that we buy only tools bearing a known brand name which we can trust.

There are many tools of unknown make on the market, perhaps cheaper in price, but until we have gained some experience to help us in our selection, it is safer to buy a named make. After all, if it is unsatisfactory we can always return it to the manufacturer with our complaint.

Regardless of whether you intend to specialize in woodworking or metalworking you are sure to need tools for both purposes and it is worth while storing them separately. At the same time, firmly resolve to

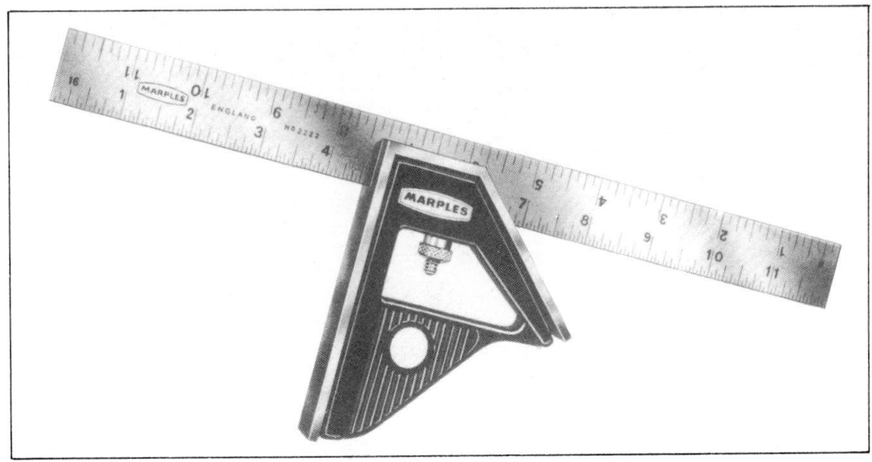

Fig. 6/1 An Adjustable Try-square, that doubles as a Mitre Gauge. A useful tool when making frames of various types.

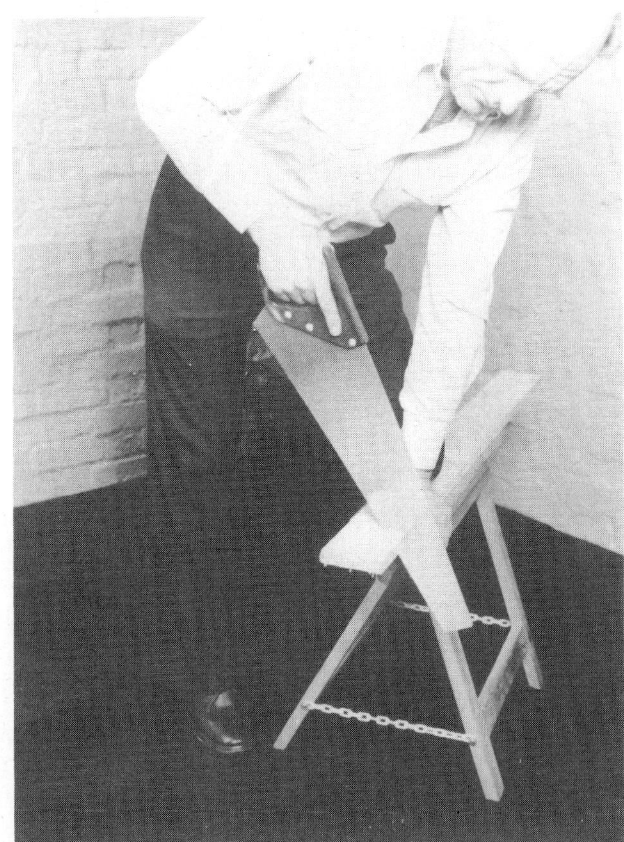

Fig. 6/2 Handsaws may be for "ripping" or "cross-cutting" (as shown here) and are named accordingly. A simple sawing trestle is also shown. Photo: Stanley Tools Ltd.

make a storage place for each tool as soon as you get it. Arrange for similar tools to "live" together, but I don't mean, for example, that all files are thrown in a drawer together!

Tools divided into groups

In general we can divide all our hand tools into various groups and, if we store them with this in mind, it becomes second nature to reach for any particular tool in the correct place and, more important, return it there.

The groups are –
1. for measuring and marking out,
2. for cutting,
3. for finishing,
4. for making holes,
5. for assembling.

Let us briefly consider each group.

1. Measuring and marking-out tools will include rulers, which had better be both for Imperial and Metric units and made preferably of stainless steel or at least, plated in some way; and trysquares, a small one for model engineering and a larger one for home woodworking. Scribers for marking metal and even the humble pencil are included.

Dividers for scribing out arcs and perhaps an adjustable trysquare will be bought as the need arises. Punches of various types can often be made. If our interest evolves to precision metal-working as in

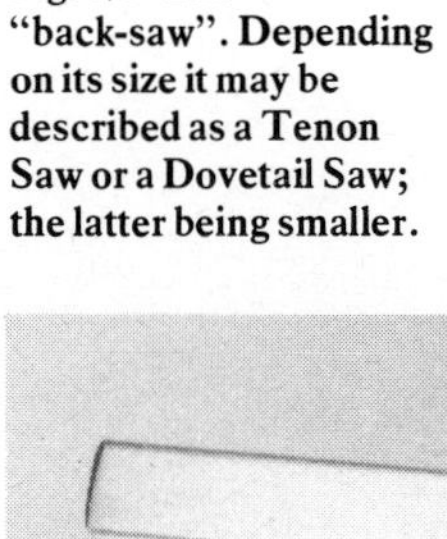

Fig. 6/3 The "back-saw". Depending on its size it may be described as a Tenon Saw or a Dovetail Saw; the latter being smaller.

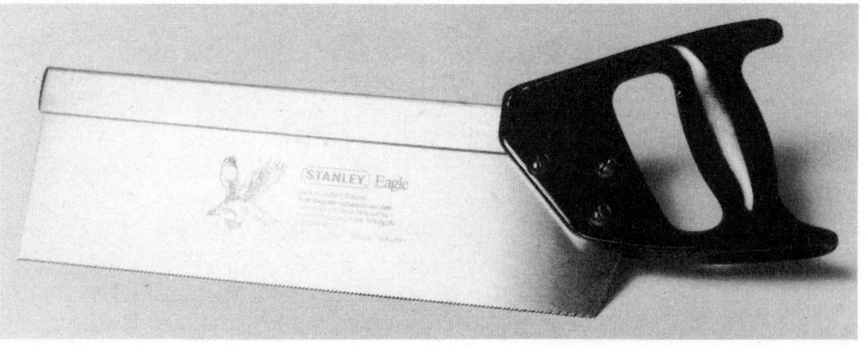

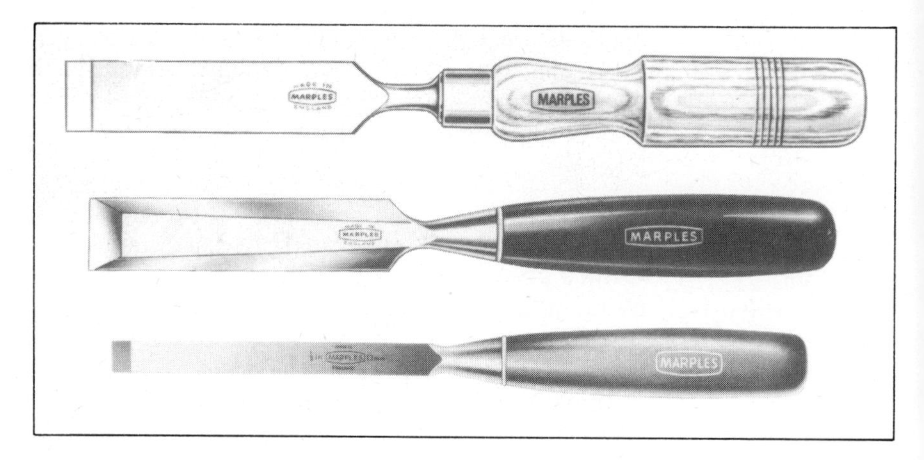

Fig. 6/4 Three types of chisel used when working wood.
Top: A FIRMER Chisel – the common chisel of the woodworker.
Centre: A BEVEL EDGE Chisel – for paring in corners when cutting joints, etc.
Bottom: A MORTISE Chisel – for cutting the mortise hole of a mortise and tenon joint.

model engineering, we may find the need for expensive tools like vernier calipers and micrometers – all useful suggestions for birthday presents! But a word of care – do not rush into buying expensive tools just for the sake of having them. Make sure that the amount of work you will do warrants the purchase of such equipment. Think about it carefully, then if you are sure that it does, buy the best.

These tools, because of their usual bright steel finish (and their expense) should be kept under cover. A drawer or cupboard (with a door) forms suitable accommodation and don't forget, as mentioned before, the VPI paper. A glass-fronted cupboard containing such tools looks nice and has the added advantage that you can easily check on their condition.

2. Cutting tools are principally saws, but we can include chisels for both wood and metal among these. Make some effort to learn how to sharpen them; when keen and correctly set their use becomes a pleasure. They are all, with a little practice, simple to sharpen, which if done regularly, takes only a few minutes.

If woodwork is the theme, a tenon saw for cutting joints plus a hand cross-cut saw should suffice. If considerable cutting down of boards is likely, invest in a hand rip-saw. However, if this is to be the case,

Fig. 6/5 A shaping and smoothing tool for wood or metal, which has a double sided blade giving a choice of straight or curved teeth and a two-way handle which changes in seconds to either place or file position. The AVEN TRIMMATOOL.

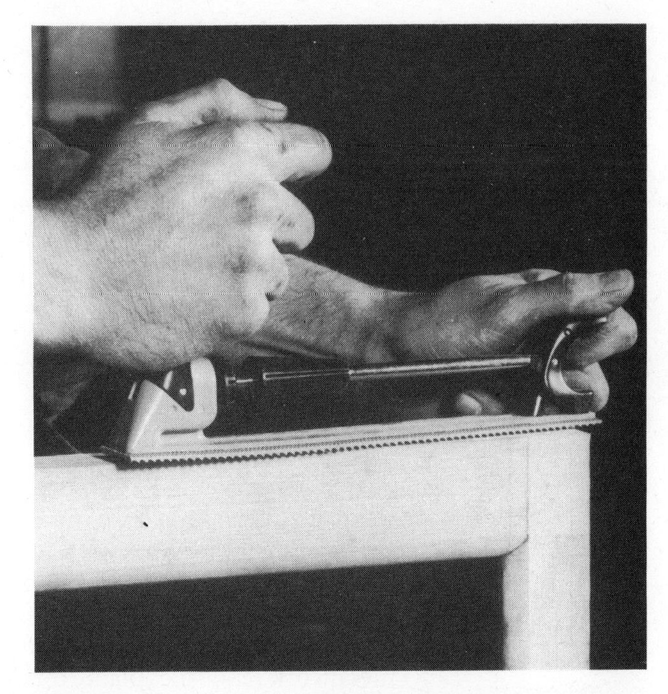

Fig. 6/6 Coarse-toothed "files" known as Rasps are useful for awkward woodworking jobs.

Fig. 6/7 Simple, but very useful hand-drill.

you may well consider setting up a motorised circular saw for the purpose.

For cutting metal a hacksaw frame is necessary, adjustable to take blades up to 12 inches long. For fine work a junior hacksaw is ideal. Note that we do not normally attempt the sharpening of hacksaw blades!

3. For finishing we would consider files for metalwork and the plane for woodworking. The former are legion in their size, shape and type of "cut" or tooth arrangement. Purchase them as needed, probably no more than half-a-dozen will cover all your requirements even if you do a fair range of metalworking. But always, without fail, buy and fit a handle to the file. Never use a file without a handle!

Although simple tools, files respond to careful treatment so store them separately in racks and if possible use a new file on the softer metals like brass, then as it dulls transfer to steel and finally let it end its days in cleaning up or fettling iron castings. This ideal situation may well not be possible in your little workshop, but it is worth bearing in mind especially if you happen to get hold of some old files.

By the way, when you have decided that the file is worn out, don't throw it away. It is a valuable source of excellent tool steel and may be made into all kinds of tools, especially if you decide to try a bit of simple forgework.

Store wood planes on their sides so that the blades remain undamaged. If they really are wooden planes, treat them occasionally with a drop of linseed oil well rubbed in, although the modern suggestion is to protect them with a coat of varnish.

There are a number of modern tools for reducing wood to size and they are all very useful but there is nothing quite like the whistle of a shaving leaving a well sharpened plane. Keep a dog-end of candle with your planes to rub on the sole of the plane when you are using it.

4. Someone said that engineering seemed to consist of producing holes then making things to fit in them! When you think of it, that is a pretty accurate description.

Most of our drilling requirements will be covered by the normal chuck-held twist drills referred to as "jobbers" drills. These are available made from carbon steel or high speed steel (HSS). The latter, although more expensive, are to be preferred especially if, as seems likely, we use an electric hand drill. These portable drills tend to run rather fast for operating on metal with the larger sizes of drill and

Fig. 6/8 The traditional carpenter's brace for larger holes to suit wood bits.

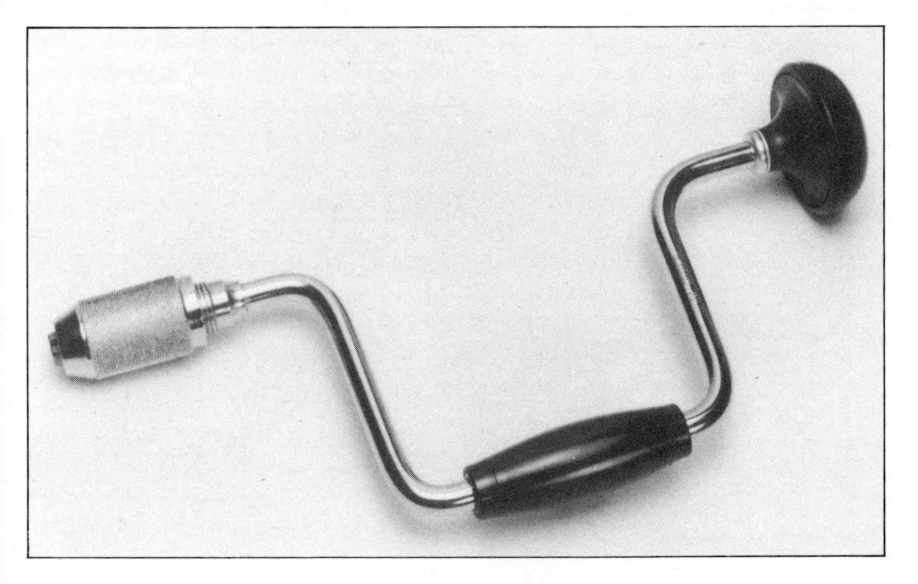

Fig. 6/9 The type of hammer commonly used for light woodworking. The narrow pein allows small nails to be started. Always keep the hammer face clean by rubbing on some fine emery cloth. A dirty face makes the hammer slip when striking the nail. Photo: Stanley Tools Ltd.

Fig. 6/10 The screwdriver is an assembling tool. Select the correct screwdriver for the screw. Drill the correct sized clearance and pilot holes for ease of entry. Using the wrong sized screwdriver will damage the screw-head, leaving metal splinters. Photo: Stanley Tools Ltd.

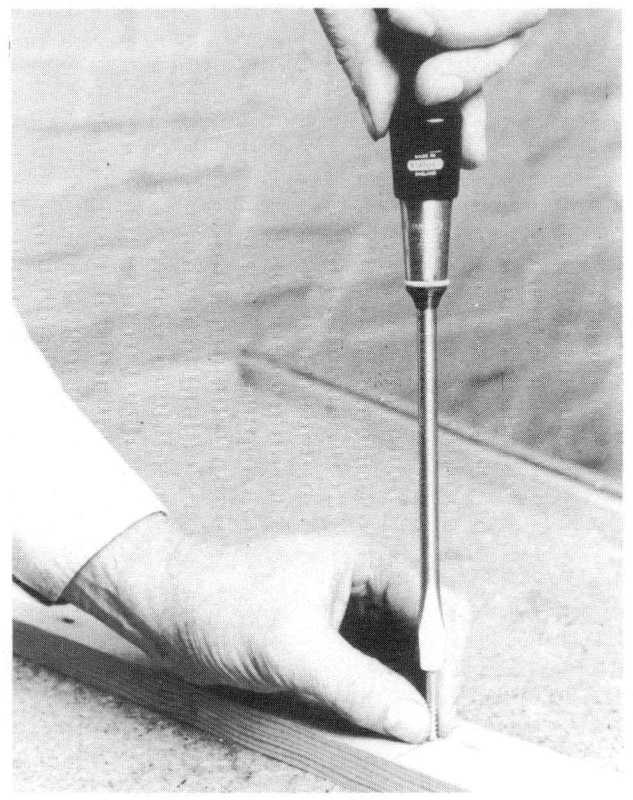

carbon steel drills would rapidly "blue" and become soft. I possess a $\frac{1}{2}$ inch capacity electric drill which runs at only 375 rpm and is absolutely marvellous. It really ploughs into the metal as though it were cheese!

Fast cutting bits are available for use in electric drills working on wood. Similarly, carbide-tipped drills can be obtained for holes in masonry.

Buy or make suitably holed stands to carry the various drills; this not only keeps them instantly available for use but serves as a reminder when they need resharpening.

Although we have talked of electric drills, the common geared hand drill, or breast drill in the larger sizes, and the old fashioned carpenter's brace are not to be despised. For many years this was all that yours truly had and for delicate work where absence of vibration is a must they are ideal.

5. Assembling covers a range of tools; soft-faced hammers and mallets, hammers with ball pein for riveting metals or claw hammers for carpentry. As your practical knowledge and skill increase so will you appreciate their special uses.

Fig. 6/11

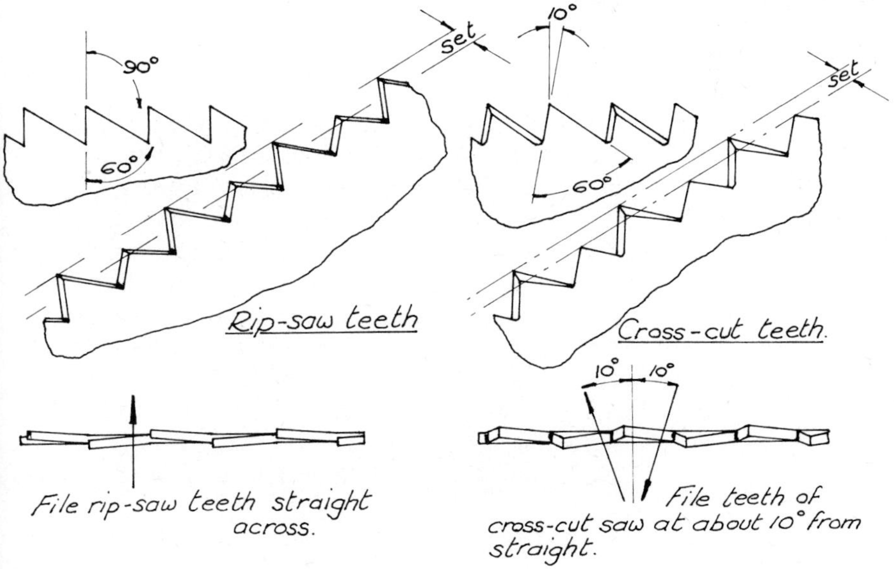

90°
60°
set
Rip-saw teeth
10°
60°
set
Cross-cut teeth.
10° 10°
File rip-saw teeth straight across.
File teeth of cross-cut saw at about 10° from straight.

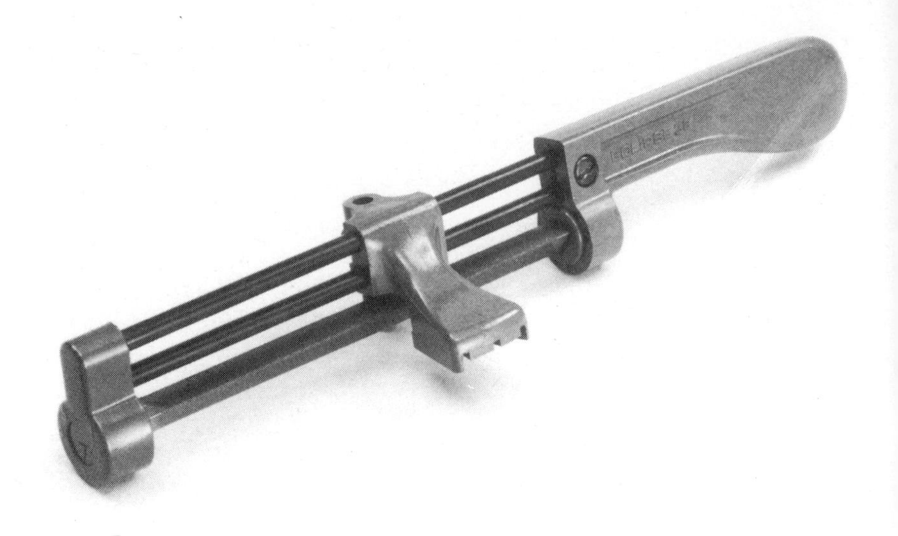

Fig. 6/12 The Eclipse Saw Sharpener offers an alternative to buying a new saw or taking the risk of sharpening without a guide.

"Handling" tools, such as the various types of pliers, screwdrivers and spanners are all in effect, an extension of our fingers, helping to accomplish operations for which we need added precision or mechanical advantage.

Wood-working tools

Although you are setting up a general workshop, the first tools you will probably require will be for woodworking, to make the bench, shelves and cupboards. It will be a poor household that can't lay hands on a saw so we will start with that.

Wood-cutting saws come in two main types divided into "hand-saws" and "backed saws". The "hand-saw" is shown in use in Fig 6.2, while a "back saw" is illustrated in Fig 6.3. As you will realise, the latter is much shorter and is stiffened by a reinforcing backbone along its top edge.

From the point of view of use, a hand-saw deals with the larger aspects of wood-cutting, while a back-saw handles the finer details of cutting joints, etc.

Dealing with the hand-saw first; it comes in a

range of sizes, usually 22 to 26 inches long and in two forms. These two forms are, first, for ripping wood longitudinally, in which case it is called a "rip-saw"; and second, for cutting across the grain of the wood (as being demonstrated in Fig 6.2), in which case it is called a "cross-cut" saw.

Reference to Fig 6.11, will give you some idea of the shape of teeth found on the two different types of handsaw. Notice that the teeth of the ripsaw are like a series of small chisels, each one cutting out a small groove along the grain of the wood. On the other hand, the crosscut saw has teeth shaped to have sharp points that scribe two parallel lines severing the grain. The remainder of the tooth then removes the waste wood from between these two lines as sawdust.

Notice also that each tooth is bent or "set" alternately to the right and left so that the saw-cut or

Fig. 6/13 The ECLIPSE Saw Sharpener in use. The saw is held in the vice clamped between two lengths of wood.

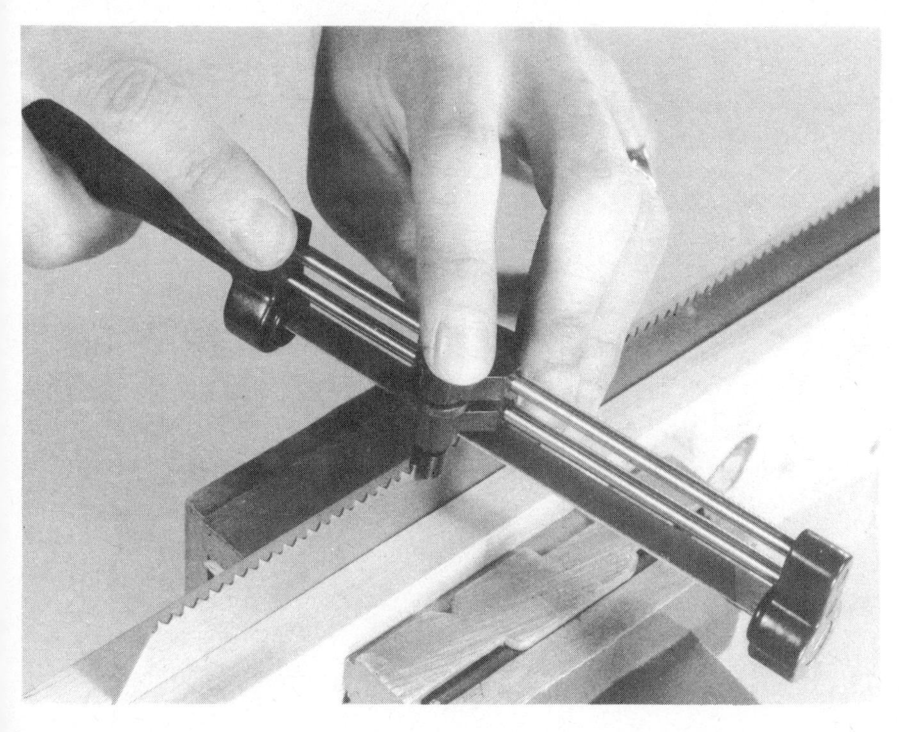

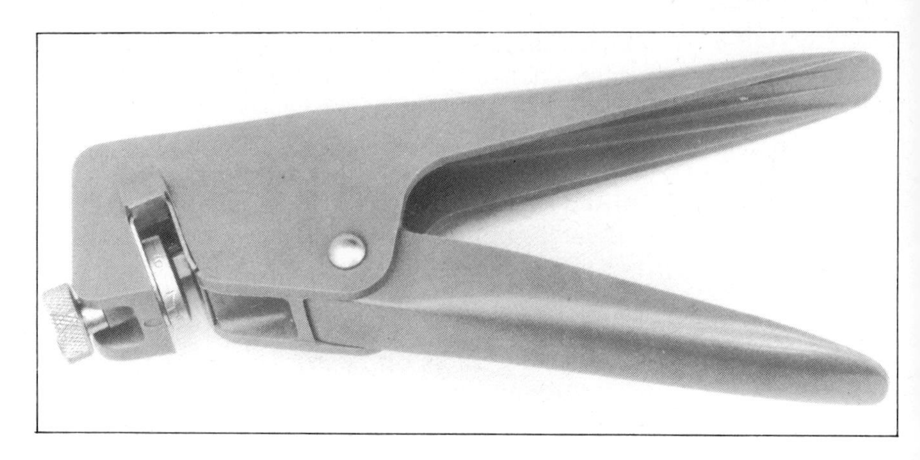

Fig. 6/14 The Eclipse Saw Set will re-set most wood saws.

Fig. 6/15 The Saw Set in use.

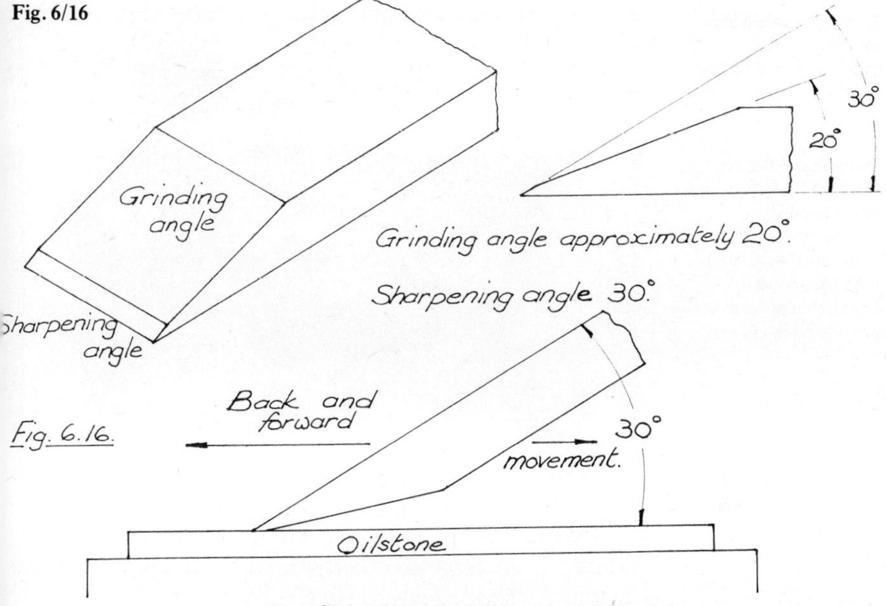

"kerf" is slightly wider than the thickness of the saw blade, thus allowing free cutting.

If you have raked out the old household saw from the garage or garden shed, it may well be in a rather dilapidated state and will need some renovation. The easy part of this will be to clean it (use fine emery cloth to remove rust), and tighten the screws holding the handle. The less easy part is to re-sharpen it, but with care this can be done and you'll have a saw as good as new.

First look carefully at the teeth and compare them with the shapes in Fig 6.11, deciding whether it is "rip" or "cross-cut". By the way, the relative coarseness or fineness of saw teeth is designated by quoting the number of "points per inch". This figure is often stamped at the back end of the blade.

The sharpening process is carried out with a double-ended file of triangular section. If you are buying one, get it 7 inches (175 mm) long. This will have a 6 mm ($\frac{1}{4}$ inch) side which should be large enough for any saw you may have to sharpen.

Clamp the saw between a couple of lengths of

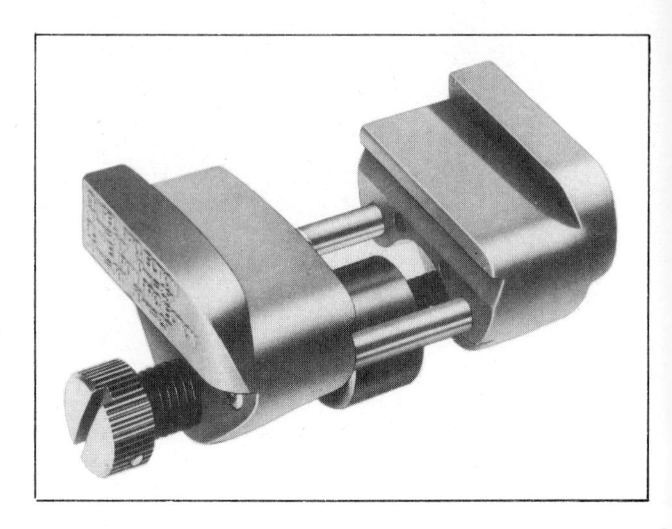

Fig. 6/17 The Eclipse Honing Guide. Designed to hold firm both wood chisels and plane irons up to 65 mm (2⁵/₈″). The blunt tool is located and secured in the Honing Guide and when used in conjunction with an oil stone, will achieve a true square edge consistently even on the narrowest of chisels.

wood in the vice, with the teeth upwards. Lay the file between the first two teeth and at such an angle that it will reproduce correctly the tooth surfaces. Make say, six steady strokes of the file, carefully keeping it at the correct alignment. Now go to the next but one space between the teeth and repeat the performance with the file held at exactly the same angle. Continue along the length of the saw taking an equal number of strokes of the file between alternate pairs of teeth.

When you reach the end, reverse the saw in the vice and repeat the whole performance between the pairs of teeth omitted in the first pass. The teeth should now be sparkling like new and if you gently lay the palm of your hand on the teeth you will feel them pricking like lots of needles. If you draw blood, don't blame me!

If some of the tooth points still look dull, then a second session of filing may be necessary. It may be that the saw is in a really bad state, with the teeth very irregular in height. If this is the case it is worth "topping" the teeth, that is, rubbing a large flat file along their tops to bring them all down to an even height before starting the sharpening operation.

After sharpening the teeth, try the saw on a scrap of wood. If it cuts freely you can congratulate yourself on a useful bit of renovation. If it tends to stick, the teeth need a little more "set" to allow them to cut a wider kerf.

In my young days, when labour was cheaper than tools, we set saw teeth by laying the saw on a block of lead and, with a flat-ended punch and a light hammer, tapping each tooth over an equal amount. If you like trying traditional methods a piece of softwood could be substituted for the block of lead.

If you feel that your skill with a file is not quite up to the requirements described above, let me draw your attention to the tool illustrated in Fig 6.12. This is a saw sharpener made by ECLIPSE suitable for saws having from $4\frac{1}{2}$ to 15 points per inch, and using a replaceable 6 inch three-square (i.e. triangular sec-

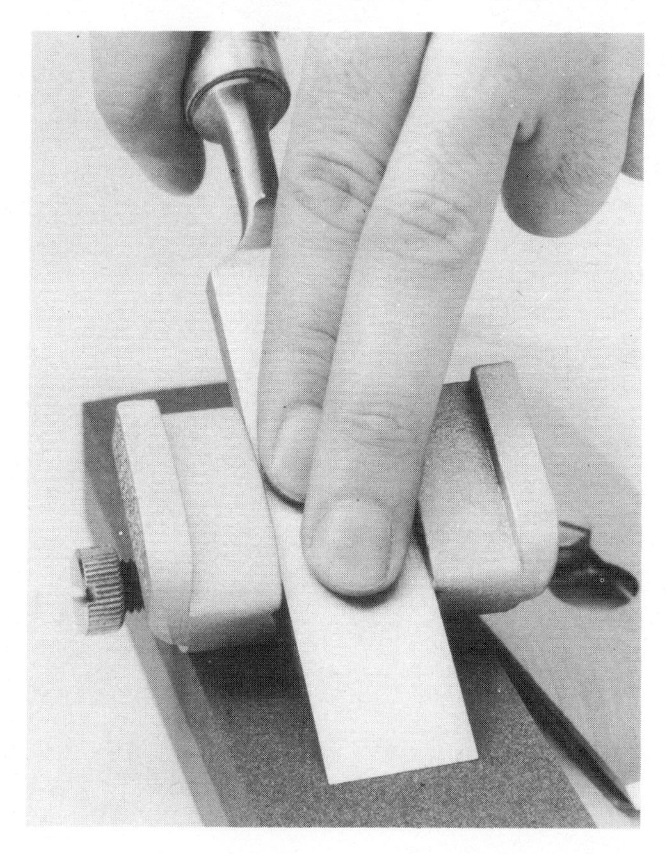

Fig. 6/18 The Eclipse Honing Guide in use.

Fig. 6/19 Although described as a Smoothing Plane, this type is commonly used by the amateur for all planing operations.

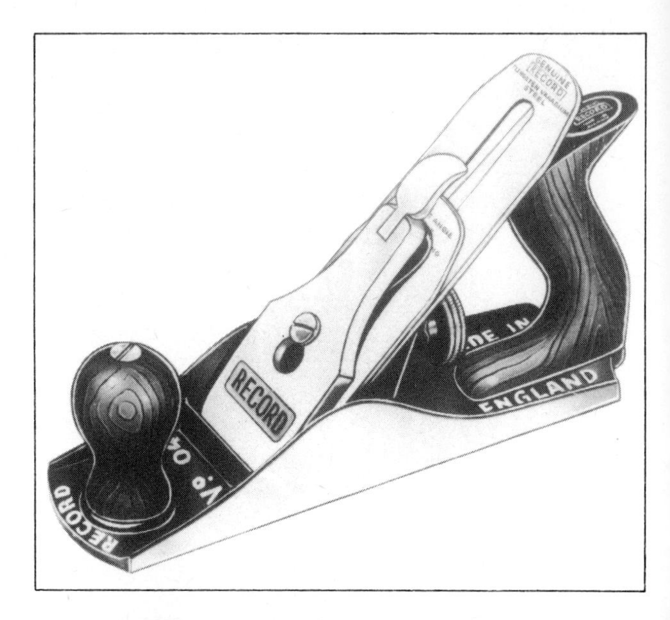

Fig. 6/20 Assembly of blade and cap-iron in a smoothing plane. Photo: Stanley Tools Ltd.

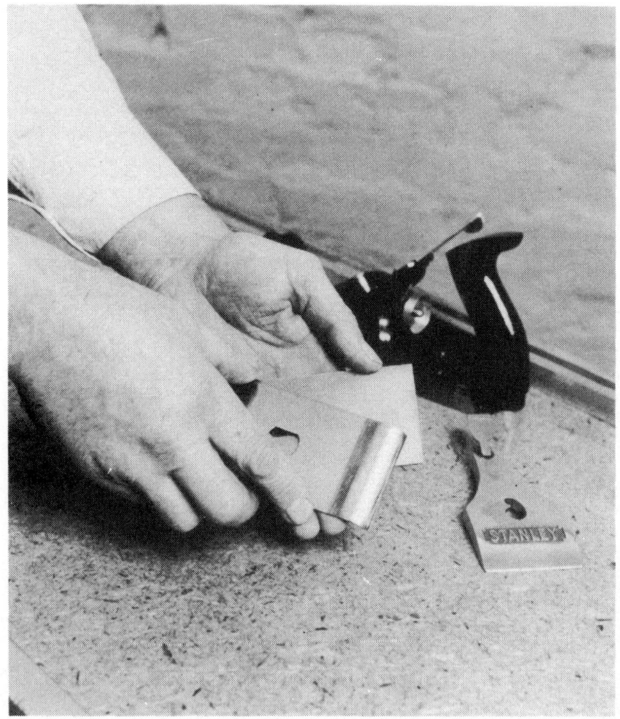

Fig. 6/21 How NOT to handle the assembly of blade and cap-iron. Photo: Stanley Tools Ltd.

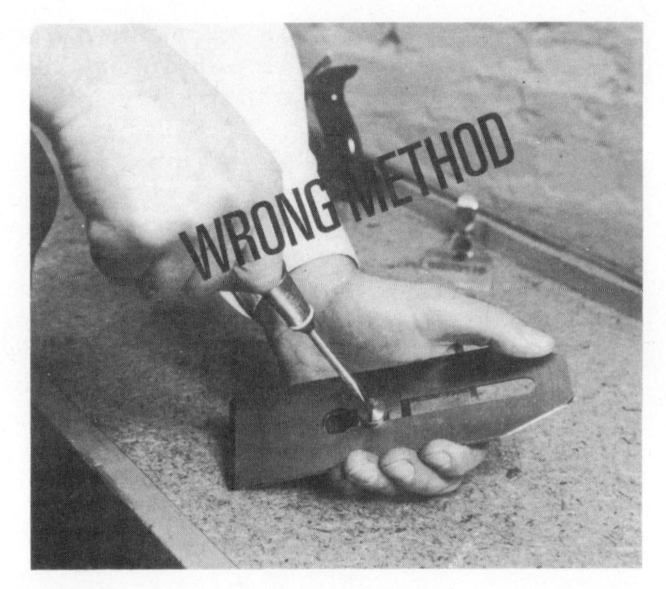

tion) file. Fig 6.13 shows the implement in use with the saw held between wooden jaws in a vice. There are five setting positions, so that various tooth profiles can be dealt with and the filing angle is automatically controlled. All you have to do is push the file back and forward.

To complete the picture, the plier-like tool shown in Fig 6.14, will accurately set the teeth of saws of from 4 to 12 points per inch. The process is shown in Fig 6.15.

Follow the instructions supplied with these tools carefully; they may differ slightly from those given above, and your woodsaw will regain the performance level it had when it left the manufacturer. Note however that these tools are not suitable for "hard-point" saws.

With a well sharpened saw to cut the timber for our bench, the next need will be for a plane to smooth the sawn surfaces and a chisel to cut the joints.

Taking the latter tool first, Fig 6.4 shows the three main types of chisels used in working wood and their main purpose. Scrutiny of a new chisel edge will show it to have two different slopes. These are illustrated in Fig 6.16. The larger slope, at about 20°, we refer to as the grinding angle, the smaller at about 30°, as the sharpening angle. It is the latter which actually has the cutting edge and which we need to resharpen. Hence when the chisel becomes dull, set the oilstone on the bench so that it won't slide about and smear the top with some thin oil. Then rest the cutting edge of the chisel on the oilstone tilted to an angle of 30° and carefully and steadily draw it back and forward along, as far as possible, the full length of the oilstone. Keep the angle constant all the time. If you find it difficult to maintain the correct angle, you may find the Honing Guide shown in Fig 6.17 just what you need. The demonstration photograph Fig 6.18, shows exactly how to use this useful piece of workshop equipment.

After ten or a dozen firm strokes, run your finger along the back edge of the blade where you may feel a wire edge. If so, sufficient metal has been removed,

Fig. 6/22 Undoing the slotted screw to separate cap-iron and blade prior to re-sharpening. Photo: Stanley Tools Ltd.

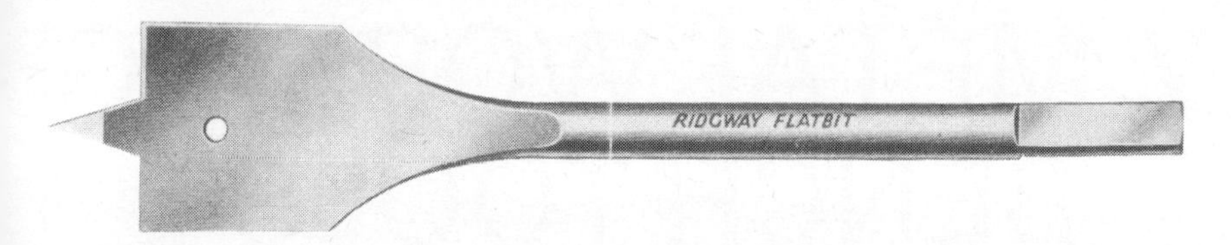

Fig. 6/23 A wood-cutting FLATBIT for use in electric drills.

so lay the back of the blade flat on the oilstone and rub back and forward three or four times to remove the wire edge.

There is an enormous range of different types and sizes of planes for woodworking. And believe it or not, some are for making the surface of the wood rougher – not smoother! However the type shown in Fig 6.19 is probably the one that will be most common in the home workshop.

Sharpening of the plane blade is exactly the same as for chisels just described and the above instructions should be followed. However, on dismantling the plane to remove the blade, it will be found to be stiffened by a "cap iron" whose purpose is to strengthen the thin blade and deflect the shavings.

A slotted screw holds blade and cap iron together and Fig 6.20 shows the arrangement. Fig 6.22 shows how the screw should be removed, NOT as in Fig 6.21.

When assembling the cap iron to the blade after the blade has been resharpened, have the edge of the cap iron not more than 1/16 inch ($1\frac{1}{2}$ mm) from the edge of the blade. If this gap is too great the plane will tend to "chatter", especially on hard woods, as the blade flexes under the force of the cut.

It remains now to consider the drilling of holes in wood. For small screw holes, etc., the common twist drills are generally used, especially if an electric pistol drill is employed. For larger holes, that is $\frac{1}{4}$ inch and above, the so-called "Flat-bits" are ideal. These must be used in electric drills as they are unsuitable for working at slow speeds such as would be obtained using a hand brace. Fig 6.23 shows such a bit, while Fig 6.24 illustrates how simple they are to resharpen.

Fig. 6/24 Sharpening a FLATBIT. Sharpening is carried out using a dead smooth file or slipstone. Take care to ensure that the original angles are maintained and that each side of the nose is sharpened equally. Do not file the sides of the blade or the boring diameter will be reduced.

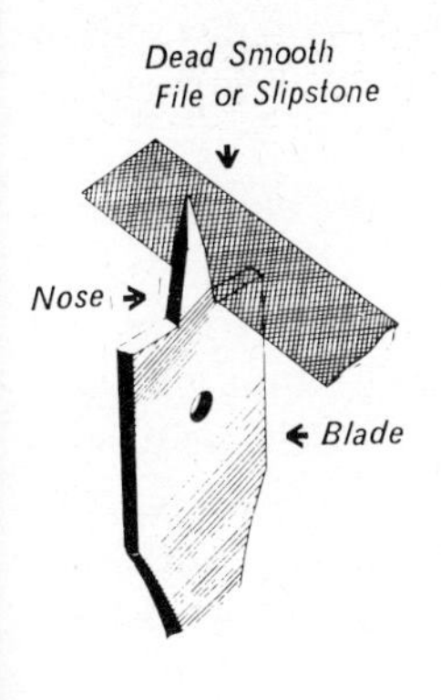

7. METAL WORKING BENCH TOOLS

We will deal in this chapter with the simple metalworking tools most useful to the reader developing a workshop.

The first thing you may have noticed is the different manner of marking out pieces of metal for cutting, filing and so on. The scaley, shiney or oily surface that we find on a metal does not lend itself to taking or retaining a pencil line; although felt pens are useful for rough marking out.

Normally a scriber is used to scratch a mark on the metal which will show up as a bright line, easily seen in good light. Although commercially available, scribers may be quickly made to the scheme shown in the sketch Fig 7.2. Drill the end of the rod to accept the point and press it in. Years ago I made such scribers from the old-fashioned gramophone needles and they were marvellous.

A brief explanation about the stuff called "silver steel" might not be amiss at this point. It has nothing whatever to do with "silver", being actually a high carbon steel which can be hardened and tempered to provide, in this particular case, a sharp point. In other examples it may be used to make cutting tools of various types.

The necessary heat-treatment is, first, heat the "silver" steel to red heat then quench rapidly in cold water. It may be that for your scriber this is all the treatment required and will give you an exceptionally hard point that may be ground and oilstoned down to needle fineness. If the point is too brittle and

Fig. 7/1(a) Fixing a mechanic's vice.

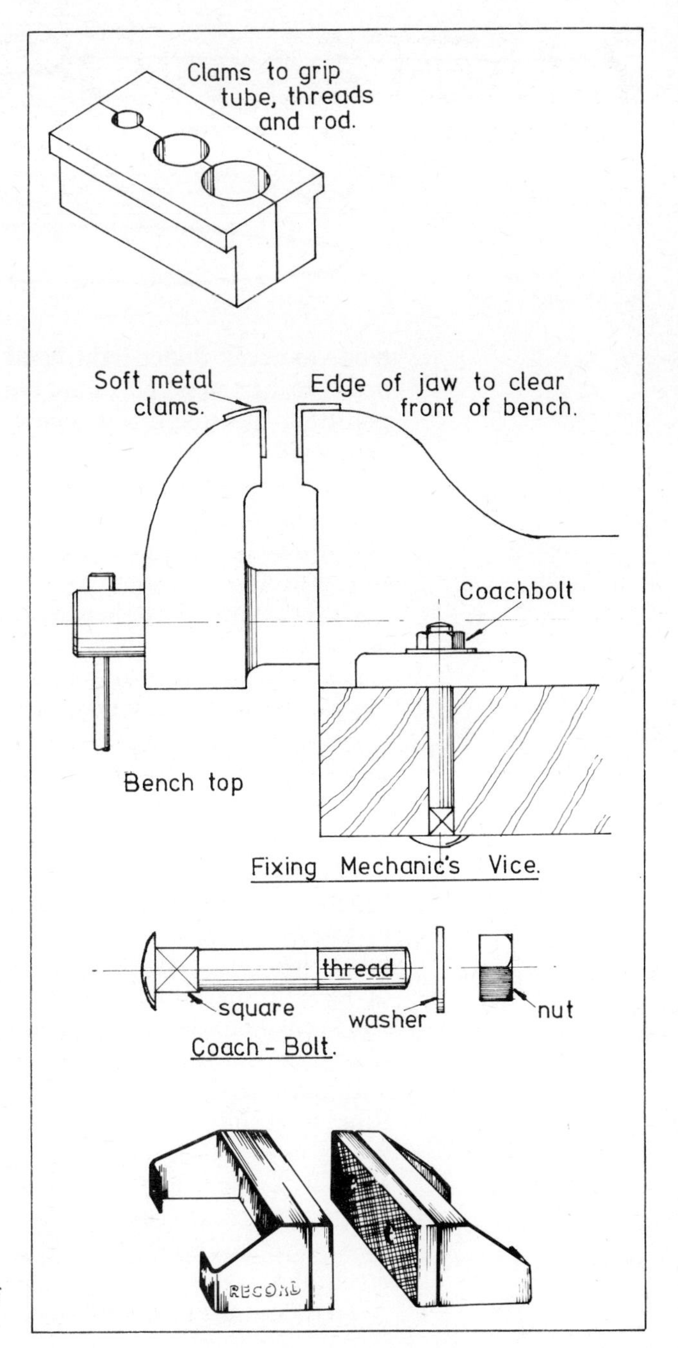

Fig. 7/1(b) Soft fibre facings protect work from being marked by the hard serrated jaws of the engineer's vice.

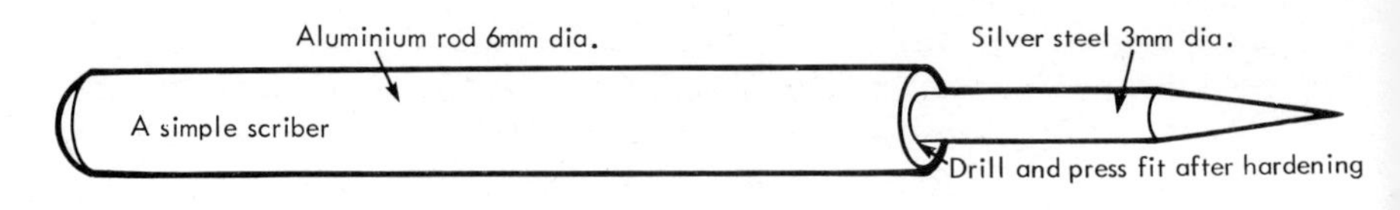

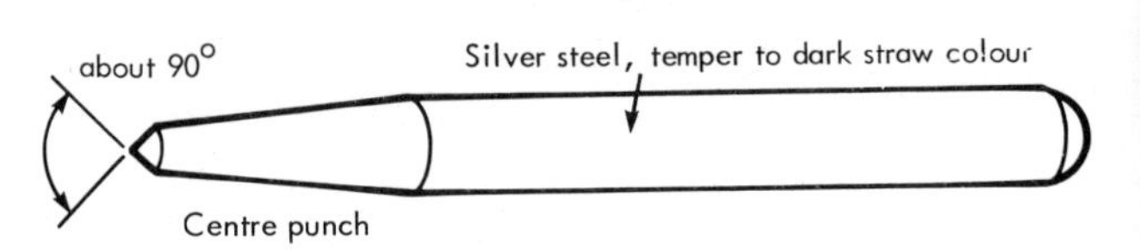

Fig. 7/2 Simple toolmaking.

tends to break under light hand pressure, the steel needs to be tempered. This is carried out by first polishing the steel bright with some emery cloth, then gently heating (really gently), until a straw coloured oxide layer begins to form on the steel's surface. This needs to be viewed in a subdued light. When the colour appears, quench the steel again in cold water.

Different degrees of hardness are required for different tools and, with care, can be accomplished using "silver" steel and these simple heat treatments. About this interesting activity, more anon.

Along with the scriber we will need a steel rule for measuring and a try-square to get things at right angles. The former would ideally be 150 mm (6 inches) long and the latter of 100 mm (4 inch) size for the kind of work envisaged in the model mechanic's workshop. Note that the size of a try-square is measured along the inside edge of the blade, the inside leg, so to speak. Such other instruments as dividers to mark circles and arcs and micrometers and verniers for accurate measurement, we will deal with later and can be bought as the need arises.

In terms of simple measuring equipment, the rule fitted with sliding heads, shown in Fig 7.3, is a very useful device. It contains an excellent rule which saves us buying one separately, and the adjustable caliper-type heads allow it to be used as a simple try-square, and a caliper, reading directly in inches or millimetres.

To set out the position of a hole prior to drilling, a

centre-punch is needed. This can be made, see Fig 7.2, in a similar manner to the scriber but from silver steel of 5 to 10 mm diameter. However, there may be a limit to how much time and effort we want to put into tool making. Mind you, making your own equipment is a very praiseworthy pastime. It can be carried to excess, but every man to his own taste.

Now that we can measure and mark out the size and shape of the metal required, we must be able to cut the same. For this a metal-cutting hacksaw, Fig 7.8, is used. Note that the demonstrator in this picture is left handed. Whether you choose a file- or pistol-type handle is up to you, but the frame must be adjustable to accept different lengths of blade.

The choice of a blade is important. They are available made either in "high-speed" steel or "carbon" steel and as "all hard" or "flexible". The "hard" type, in the hands of a skilled user are highly efficient metal cutters, but the "flexible" is likely to give a considerably longer working life if the user's actions are less skilful. Whether you choose "high speed" or "carbon" may depend to some extent on the depth of your pocket, but the former have a much longer life and will cope with harder metals. It is very much a case of paying your money and taking your choice. But do use a good branded make.

Fig. 7/3 A particularly useful tool is this caliper which fills the gap between a steel rule and a precision vernier caliper.

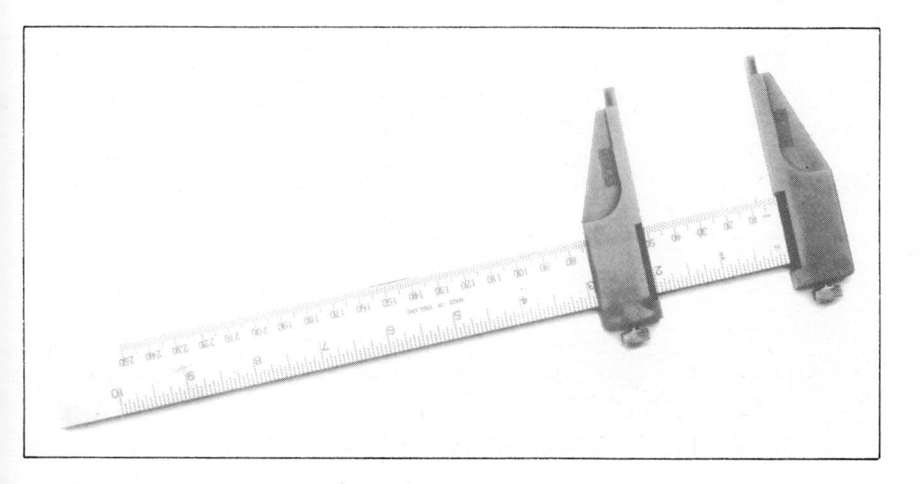

Fig. 7/4 Using a Moore and Wright Combination Set as a try-square.

Fig. 7/5 The precision protractor of the Combination Set being used to set angular work.

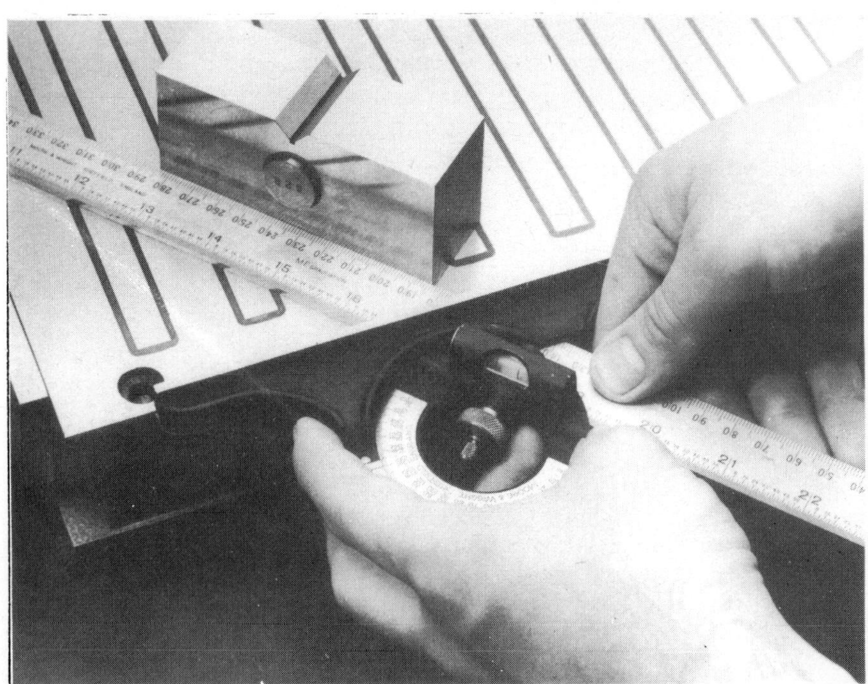

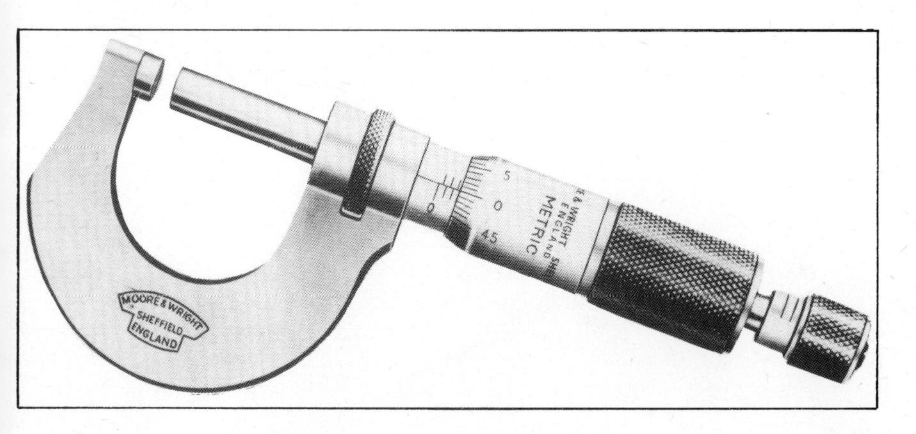

Fig. 7/6 If your interests develop towards model engineering or precise metalworking a micrometer may become necessary. Books in the Argus range explain the purpose and use of such instruments.

In addition you will need to choose a particular tooth size, or number of teeth per inch. Generally 18, 24, and 32 teeth per inch are standard, although others are available. For cutting steel and cast iron use 24 tpi, for aluminium and copper alloys the coarser 18 tpi will reduce clogging, while for sheet and tube 32 tpi will be about right. If choosing one grade, go for the middle one.

Breaking of hacksaw blades is generally the result of them being twisted or wrenched in the cut. Loss or rapid wear of the teeth is invariably due to them straddling thin work with the result that the tips of the teeth are broken off. There should always be at least three teeth in contact with the metal in the cut.

Two other types of saw will be useful in the workshop. For fine work, the so-called "Junior" hacksaw, Fig 7.9, consisting of a simple frame with a miniature blade is ideal; while to use up the blades inadvertently broken, we can obtain a simple handle into which the broken remains can be clamped for those occasional awkward-to-get-at places.

The ability to hacksaw accurately is a valuable skill and when learnt will save hours of hand filing. Which seems to bring us to the next metalworking tool in our mechanic's workshop – the file.

These come in a wide range of shapes. Flat, square, round, triangular and tapered section and in sizes from about 16 inches long downwards. It is

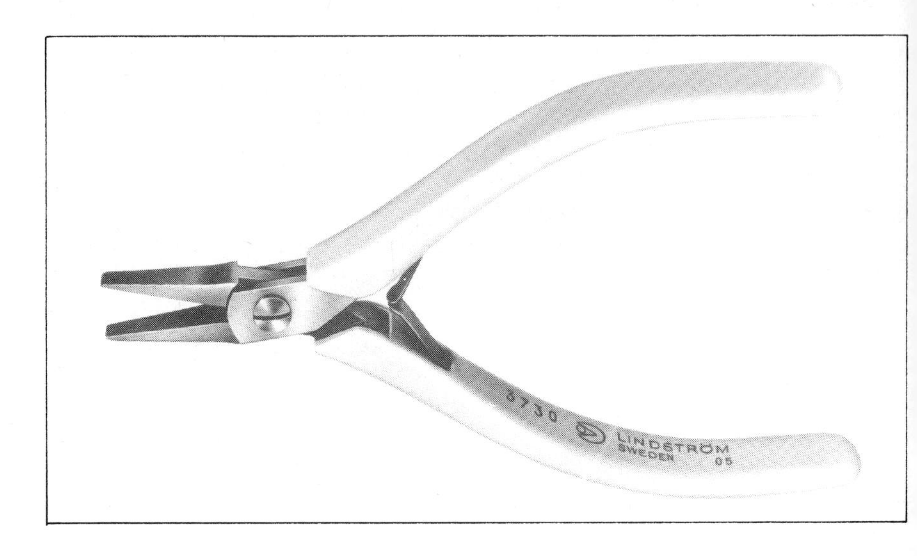

Fig. 7/7 Pliers come in an enormous range of types and sizes. The flat-nosed type shown are very useful to the model engineer as they do not mark delicate work.

rarely that we will need any greater than 8 inches long and as a start two or three of this size should suit us for the odd jobs about the workshop and house that will give us some practice in their use.

A useful pack, Fig 7.10, can be bought and we can gain experience in handling them by renovating a few pieces of gardening equipment. Figures 7.11 (a & b) illustrate the method of handling files. If you happen to be left handed, hold the pictures up to a mirror!

Drills and drilling

It has been said that engineering consists of making holes and putting things in them! This is, in fact, not a bad description and illustrates how important the drilling of holes is in the Home Workshop. In the last chapter I touched on the subject and because it is such a necessary part of working in metal, I propose expanding the topic further.

First, the drills themselves. For drilling metal these are certain to be of the type described as "twist drills", and are available in an enormous range of sizes. Certainly in my own experience I have used them from No 80 (0.0135″) diameter, up to nearly 2 inches diameter. The days of the home-made, arrow-pointed drill would appear to be gone, although a very old model engineering book I was reading recently described how the amateur should make his drills from various sizes of steel knitting and sewing machine needles!

The diagram, Fig 7.14, shows the usual form of the business end of a twist drill. The point is basically conical in shape and includes an angle of 118°. Thus the important process of resharpening a dulled drill starts with offering up the drill point to the surface of a grinding wheel and rotating it so as to produce a conical end having the included angle mentioned above. As the included angle is near enough 120°, we can use a 60° setsquare to facilitate getting the drill positioned at the correct angle.

So far the process has been no different from that of regrinding say a centre punch. The snag is however, that a drill with a perfectly conical end will not have the necessary clearance to allow its two cutting

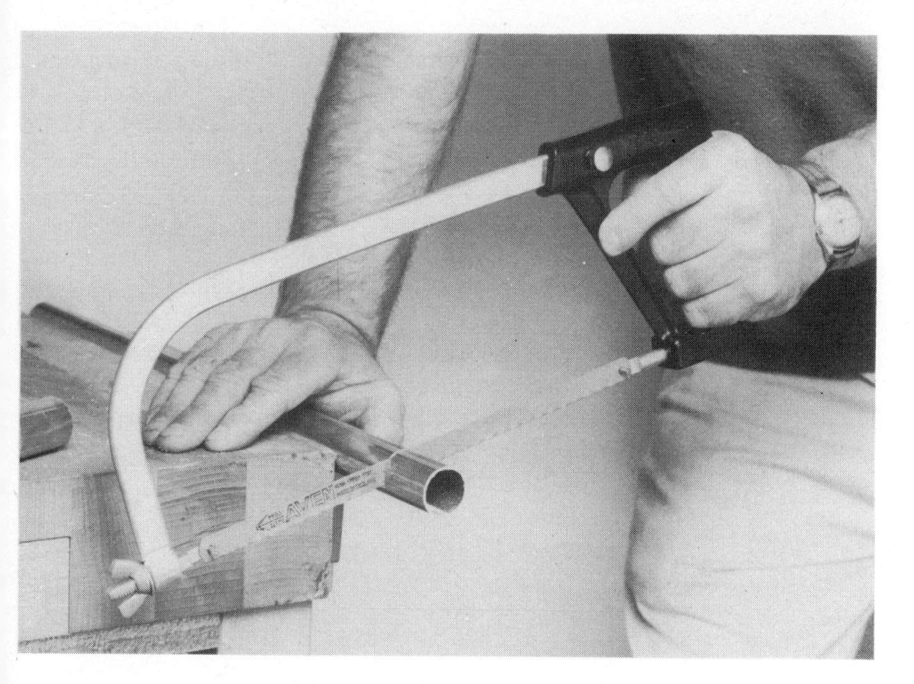

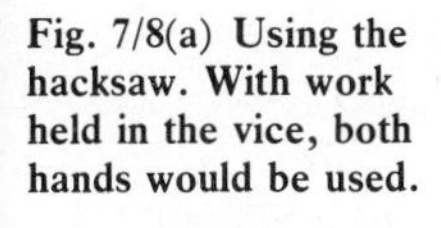

Fig. 7/8(a) Using the hacksaw. With work held in the vice, both hands would be used.

Fig. 7/8(b) A "pad-saw" handle makes use of broken hacksaw blades.

edges to penetrate into the metal during the drilling operation. Hence the conical surface behind the cutting edges must be "backed off", i.e. curved inwards, to give a clearance which will allow the cutting edges to "plough" into the work. Such a clearance should be about 12° when viewed as shown in the diagram.

Invariably any complaint of poor drilling, need for excessive pressure, or signs of overheating, will be due to there being insufficient clearance behind the cutting edge. Presuming of course that we are not trying to drill metal that is too hard for the drill.

The end of the drill does not come to a point but, due to the thickness of the central web and the disposition of the cutting edges, terminates in a

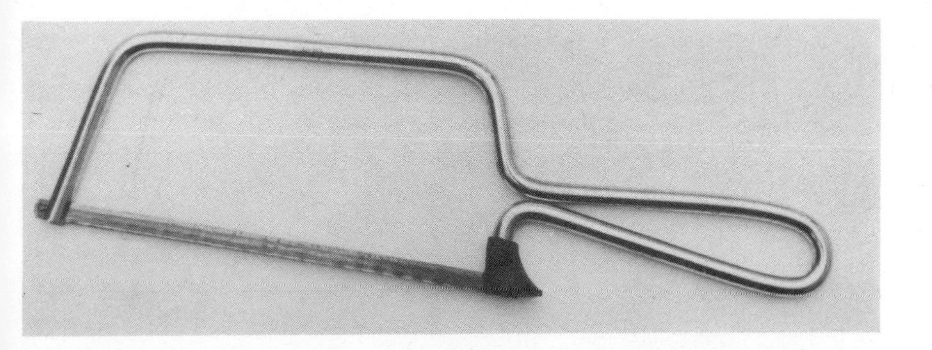

Fig. 7/9 **For fine work the JUNIOR hacksaw is ideal.**

chisel edge. This cuts purely by scraping out the metal, not exactly an efficient way of producing a hole. Though in the smaller sizes its inefficiency is not pronounced, it is advisable when drilling anything over about 6 mm ($\frac{1}{4}''$), to drill a pilot hole first.

The best way to learn how to sharpen drills freehand is to beg or borrow a twist drill of reasonable size, say about 12 mm ($\frac{1}{2}''$), which has been correctly sharpened. Offer it up to the grinding face of an abrasive wheel (stationary, of course), with the cutting edge disposed horizontally and pointing upwards. Now slowly rotate the drill in your fingers as though forming the conical end, but keeping the drill's conical surface in contact with the abrasive wheel throughout. You will soon see that to maintain contact between drill and wheel you must cause the shank end of the drill to drop slightly as it is rotated.

Repeat this action a number of times until you have got the feel and flow of the necessary finger and

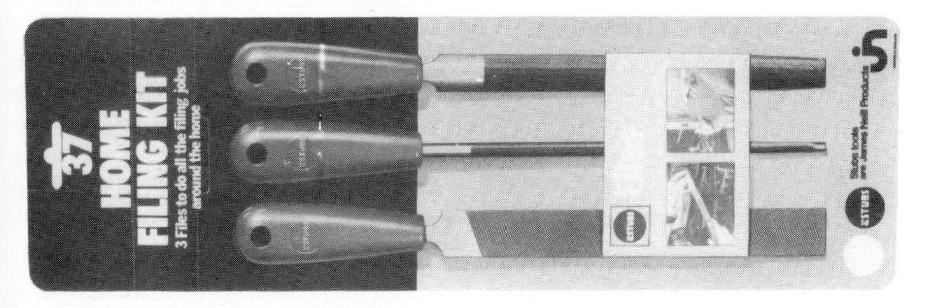

Fig. 7/10(a) **It may be a good idea to start with a handy set of files.**

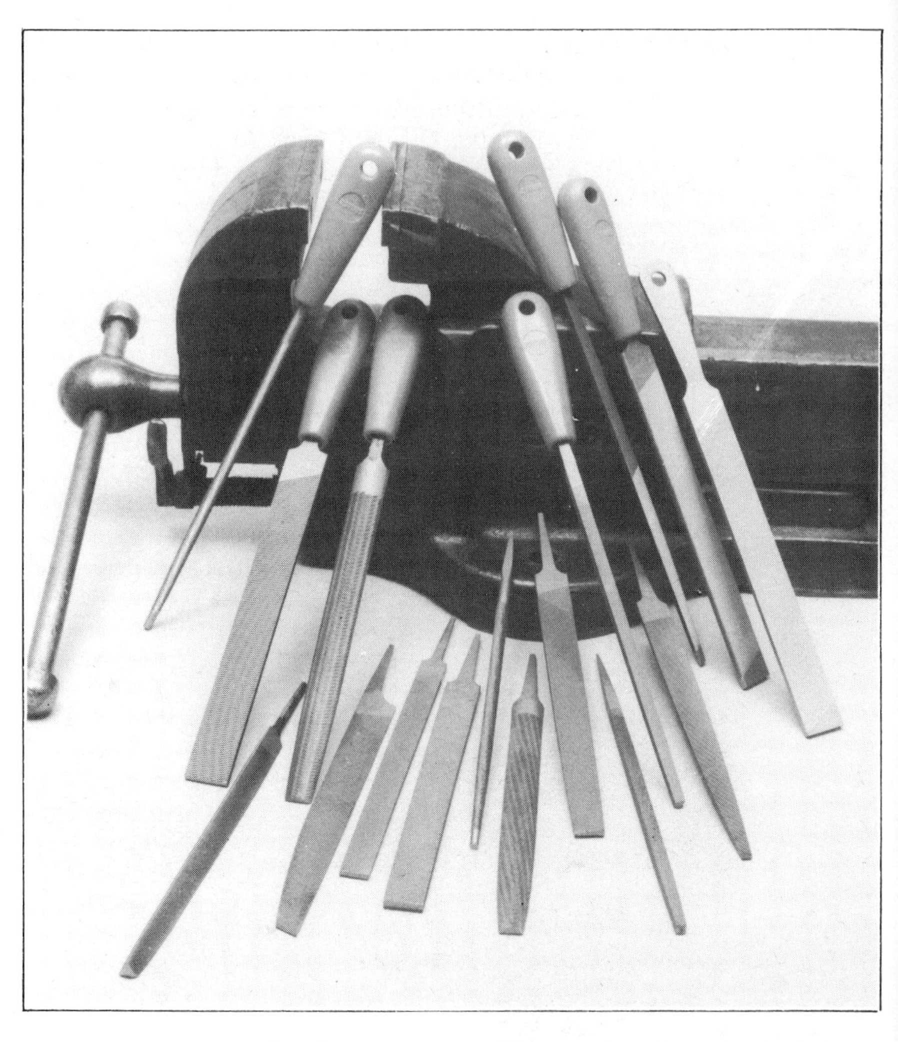

hand movements. Then, with the grinding wheel running and a blunt drill of similar size (your own this time), see if you can successfully reproduce the necessary shape. Each cutting edge needs to be treated so that when carefully sighted they are symmetrical.

With tiny drills, the "rolling" technique becomes rather awkward. A simplified method is to offer the drill up to the abrasive wheel at the required 60°

angle together with a downward inclination of about 12° to give the necessary clearance. The surface behind the cutting edge will be flat, not conical, but if used carefully the slight weakening of the cutting edge is not sufficient to cause any problems.

Alternatively for small drills, the same techniques may be applied to offering up the drill to a flat oilstone. With the cutting edge facing forward, tilt the drill to the left until it forms an angle of 60° with the horizontal, then incline back at 12°. The drill is then drawn back and forwards in a similar manner to sharpening a woodwork chisel.

Fig. 7/11(a) Files are efficient and very adaptable tools though simple. They are used in all practical situations from DIY car repairs to precision tool-making.

This is basically the principle used in the ECLIPSE drill sharpening device shown in Fig 7.15, and if you are at all doubtful of your ability to perform the operations described above you are strongly advised to get one of these useful pieces of equipment.

Regarding the use of an abrasive wheel; the plain type must be used only on its periphery as shown in Fig 7.14, never on its side. If you wish to grind on the side of the wheel you must use a cup type. I know we don't always practise what we preach – but you have been warned!

If you never break a twist drill, you can probably walk on water –! So when you do, don't throw them

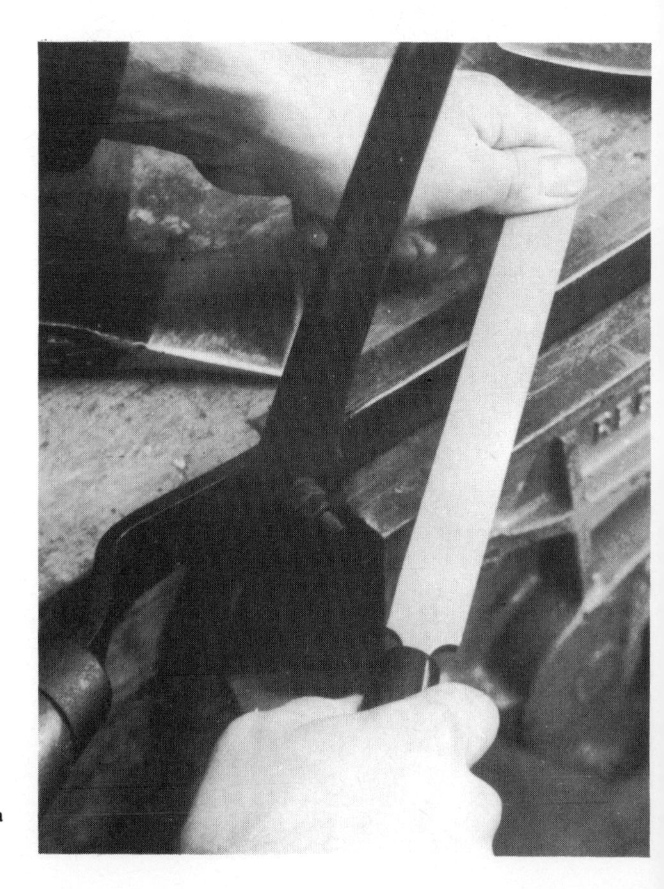

Fig. 7/11(b) Handling a file.

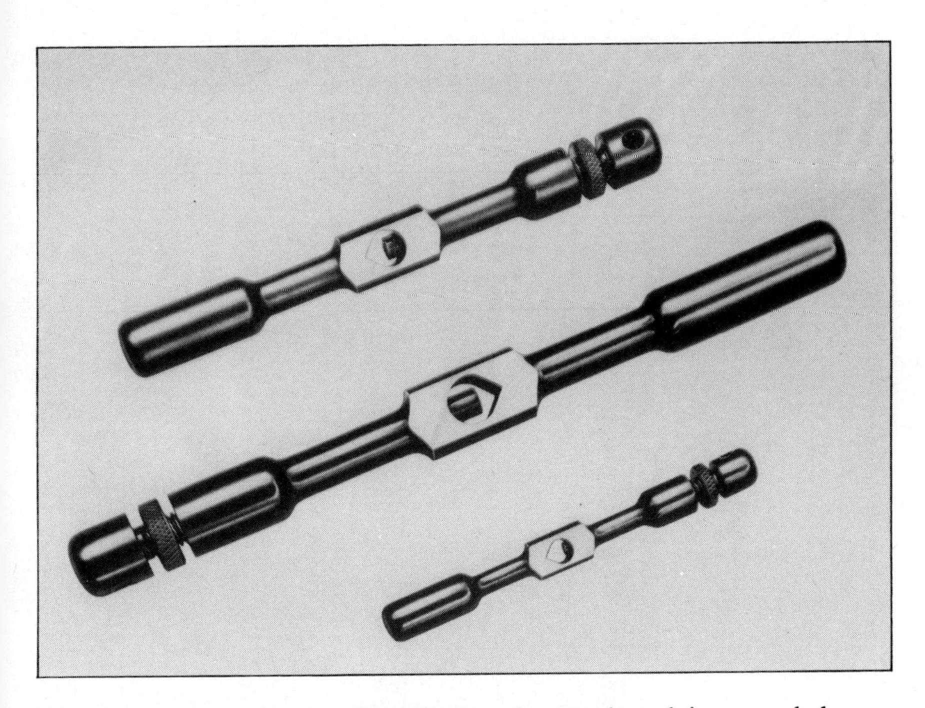

Fig. 7/12 Tap Wrenches come in a wide range of sizes. Their use is referred to in chapter 14.

away. The shank end, even though it may only have a short fluted length remaining, can be reground to make a short, stubby drill, ideal for use in the hand brace, while the front portion can be soft soldered (silver soldered if it is high speed steel) into a piece of mild steel rod and continue in use. I have heard of such recovery jobs being done with one of the modern, high strength, adhesives, but have no personal experience of this to date.

Tool sharpening comments

In the last couple of chapters we have had quite a lot to say about sharpening tools of various types and before the non-technical reader shies away from this necessary part of owning a workshop, perhaps a few comments should be made on the subject.

The possession of workshop tools of various types is simply a nuisance if we do not know how to maintain them in good condition and sharp. There is no need for the sharpening of simple tools to be any

AVEN
HALF ROUND
FILEMASTER
HALF ROUND
Example of curvature
A PROFESSIONAL TOOL FOR THE HANDYMAN
AVEN
FILEMASTE
AVEN
HALF ROUND
FILEMASTER

more than learning a simple technique and then careful practice.

The technique may be most easily appreciated by relating it to the equipment used for sharpening cutting tools of various types, such equipment consisting of the oilstone, the file and grinding wheel.

Oilstone

The oilstone we all recognise as an oblong slab usually protected by being housed in a wooden box. It is so called because thin oil is used to lubricate the sharpening action and to keep the "stone" from becoming loaded with the steel dust from the sharpening process.

Traditionally it is the means of sharpening woodwork chisels and plane blades, but lathe owners often use it to give a keen edge to their turning tools and with a simply made device we can resharpen small twist drills on it.

The basic technique as described in the previous chapter is used when sharpening any article, whether a simple penknife, metalworking chisel, lathe tool, etc., on an oilstone, once you have determined the correct sharpening angle.

Files

For saw sharpening, files are used and special double ended ones of triangular section are made for this purpose. Woodworking saws are of two types, for cutting across the line of the grain, which are the saws you will find most useful, and for cutting along the grain (or ripping) which you will only use if you often have to reduce the width of planks of wood; again refer to Chapter 6 where the "modus operandi" of saw sharpening is explained.

Using files we can often sharpen metal cutting chisels or "cold chisels" as they are generally called. They get that name to differentiate them from the "hot-chisel" used by the blacksmith to cut iron at red heat on the anvil. The simple sharpening process appears in Fig 7.16.

Fig. 7/13 Manufacturers and retailers have the needs of the amateur user in mind in the way they market tools and accessories.

Also many garden and household implements can be sharpened with a file after a careful study of the shape of the blade and the application of that modicum of common sense possessed by all "workshoppers".

Grinding wheel If you spend a lot of time in your workshop, it will not be long before you feel the need of a grinding wheel for your tools. Even if you are concerned

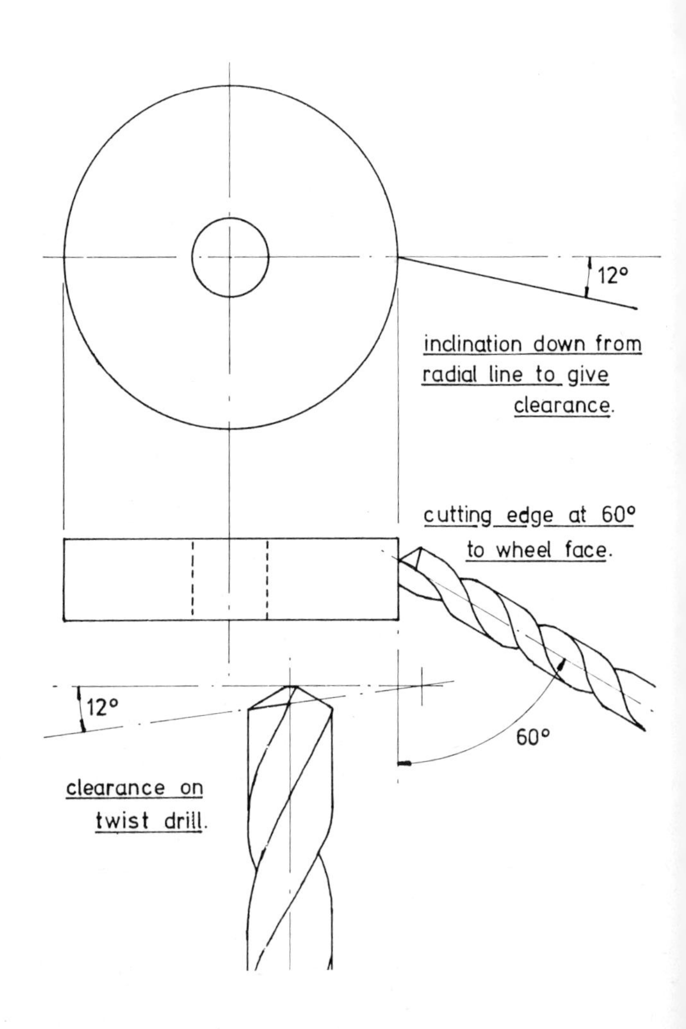

Fig. 7/14

Fig. 7/15 The ECLIPSE twist drill sharpener.

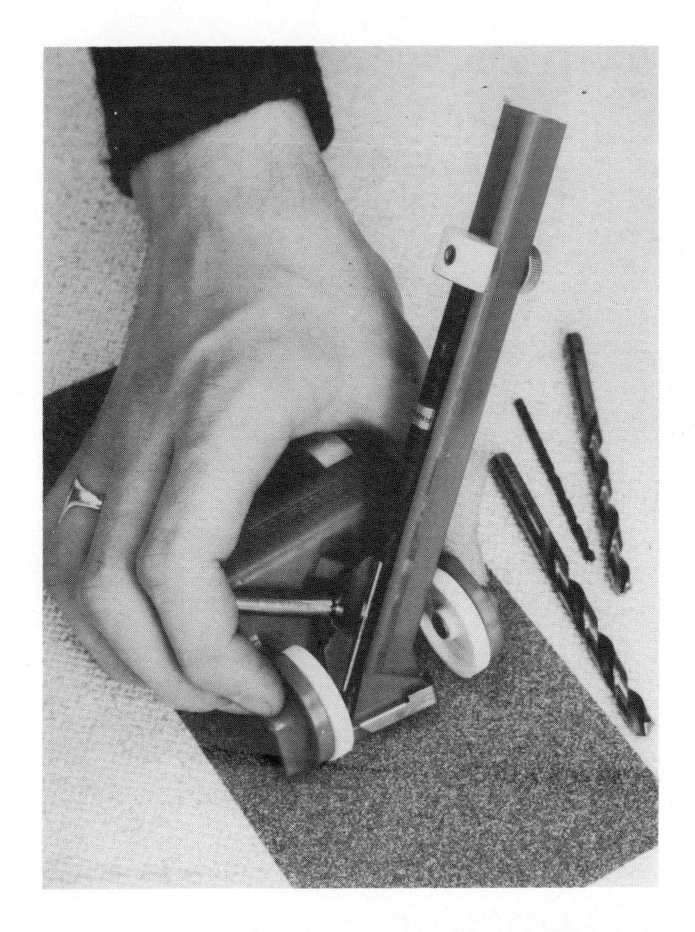

mainly with woodworking, you will soon find that resharpening of plane blades and chisels has removed most of the 20° grinding angle and on each resharpening you are having to take off a greater amount of metal in order to get a keen edge. The time has come to re-form the grinding angle, and, if this is not going to take a long time, an abrasive wheel has to be used.

There are many different commercial versions of tool grinders available to the amateur depending on the depth of his pocket, from the simple, cheap but very useful polishing and grinding heads to the double-ended motorised type complete with tilting tables and coolant supply. Whichever you choose or build, let it be a substantial machine, running at correct speed for the abrasive wheel size with efficient guards and tool rests.

In industry, Tom, Dick or Harry is not allowed to be responsible for the setting up of abrasive wheels unless he has been on a special course dealing with the techniques and safety requirements of this very critical operation. These conditions are not legally enforced in the home workshop but that is no reason why you should not, in your own interest, put them into practice. The "Abrasive Wheels Course" usually of one day duration, is run at many local technical colleges and will teach you a lot. Don't let it worry you that you have only one tiny grinding wheel and you are sitting with men responsible for hundreds – if yours bursts it will be just as lethal!

If you set up a small grinder, here are some points to note. The wheels must be effectively guarded with a metal guard (probably steel) which fits the wheel fairly closely. The tool rest must always be set close to the wheel so that nothing can be trapped between rest and wheel. Don't hold work with rags or wear gloves "because it is getting hot"! The maximum running speed for the type of wheel you will be using

Fig. 7/16 Sharpening tin-snips and cold chisels.

Sharpening

SNIPS AND SHEARS

Open the jaws as far as possible and present the relieved cutting edge of each blade to the grinding wheel (preferably an 80 Grit bonded wheel) at an angle of 80°. Starting at the joint, draw the blade across the grinding wheel.

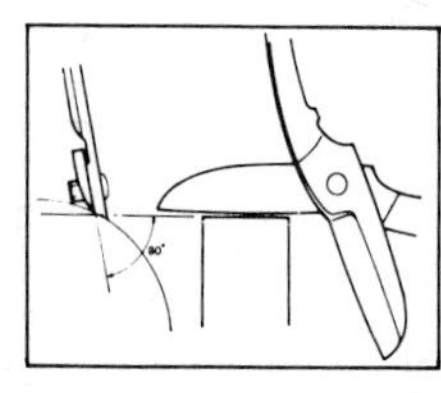

To prevent damage to the blade, a grinding coolant should be used (water or soluble oil) and the amount of steel removed should only be sufficient to clean up the cutting edge. The grinding process may create a burr on the cutting edge and this MUST be removed on a smooth oil stone. On no account should the inside face of the blade be ground.

At all times ensure that the pivot bolt is tight and well oiled.

COLD CHISELS

All GILBOW Chisels are designed and manufactured so that the important sharpening operation can be performed with a SMOOTH FILE. It is essential that in sharpening, the angle of the cutting edge should be retained (approximately 65°).

will probably be 6,500 surface feet per minute; that is, a point on the periphery of the wheel will be passing the tool rest at that speed. Thus, should you build a grinder to be powered by the usual $\frac{1}{4}$ hp electric motor, with pulley sizes giving a 1 to 2 ratio, i.e. the motor runs at 1425 rpm and the grinding spindle at 2850 rpm, the largest safe diameter of abrasive wheel you could use would be 8 inches. Such a wheel will probably be much too large and heavy for your little grinder and probably a 5 or 6 inch diameter by $\frac{1}{2}$ or $\frac{3}{4}$ inch wide will be chosen bearing in mind that the guard should fit it closely.

8. POWER TOOLS, INSTALLING ELECTRIC MOTORS

We do not have a workshop for very long before beginning to feel that it would be nice to have some powered equipment. This may be to lighten monotonous jobs, like sawing wood or drilling holes, or it may be to open up for us a new field of skill such as in the installation of a small lathe.

Safety, confidence and care

If we have had no experience of workshop machinery, we may look upon its acquisition with some trepidation. But if we bear in mind three simple rules we need have no fear about taking advantage of powered equipment.

First is safety; this must be constantly in our mind. There are two aspects of this, first, the machine itself must be as safe as we can make it, always remembering that any machine is only one hundred per cent safe if it is (a) stationary, and (b) not being used! Second, we must work safely. For example, a correctly fitted guard on a circular saw is "machine safety", but if we try to saw the corner off a tiny piece of wood with our fingers under the guard, then we are guilty of not working safely.

The second rule is confidence. Plan the operation you intend doing on the machine. See that the cor-

Fig. 8/1 The popular electric drill.

Fig. 8/2 Electric drill with rotary sander attachment, suitable for general DIY duties.

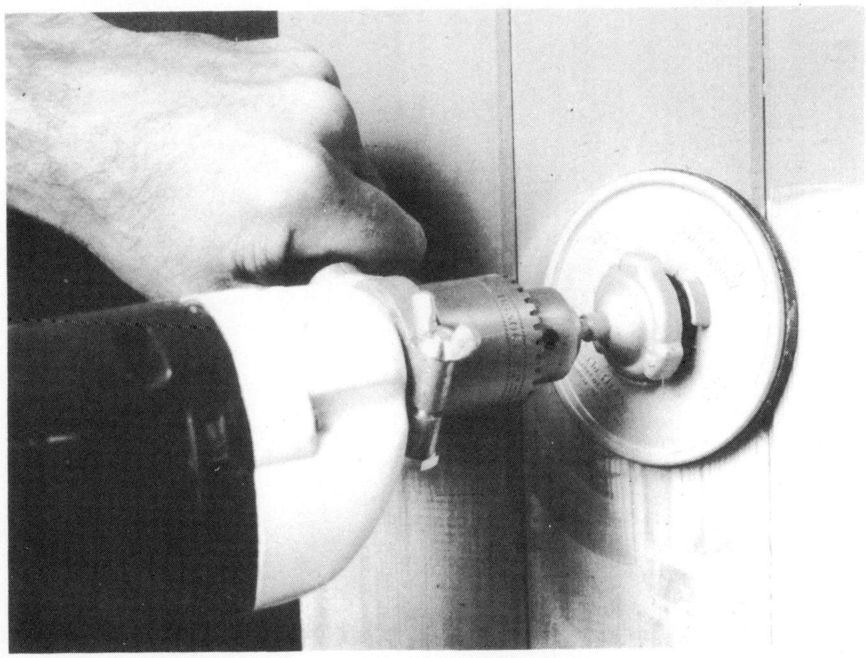

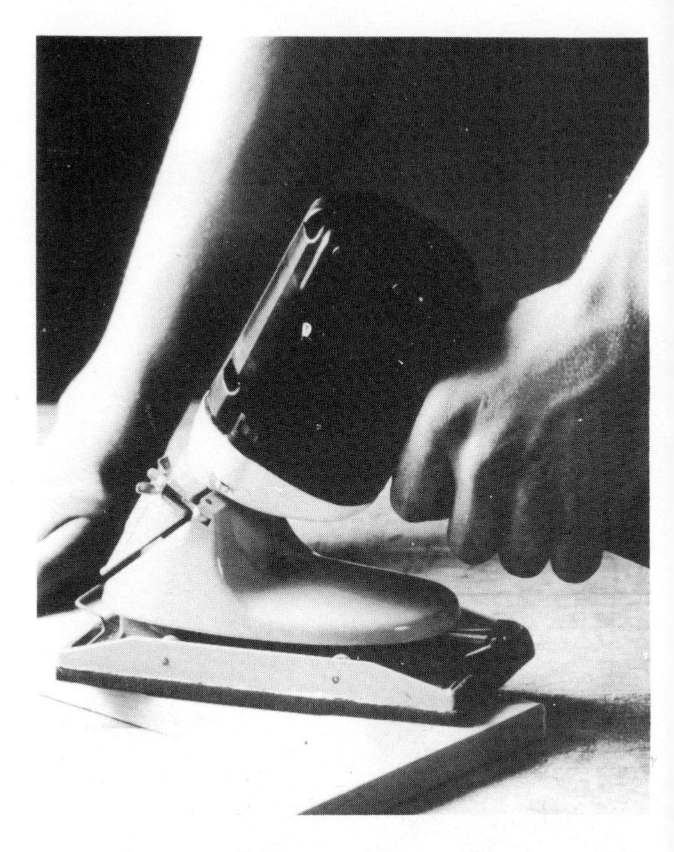

Fig. 8/3 The orbital type of sanding attachment, more suitable for fine woodworking.

rect tools are available – and sharp, that the work is firmly supported, correct speeds and feeds are used, and then confidently and with complete concentration on the job in hand, get on with it.

The final rule is care and this relates to the concentration mentioned above. Give your whole mind to the job being done. Personally I cannot operate, say a lathe and listen to the wireless at the same time; to me "music while you work" is just "not on"! If a friend calls, we both sit down and have a natter.

Portable power tools

No doubt the first power equipment you will have will be an electric drill of the portable type. Then you will start adding accessories to this for sanding and sawing wood. Perhaps an abrasive wheel for tool sharpening (don't forget to make up a simple guard for the wheel and keep the work rest correctly adjusted). We may invest in a drill stand, so that drilling holes becomes easier and their location more precise. Or we may purchase the necessary fitments to convert the portable drill into a small wood-turning lathe.

For light work such equipment is ideal, and it may be that it will satisfy your requirements completely. However, if your interests demand heavier equipment or more precise and specialised machine tools you may feel the need to graduate to larger, probably commercial tools.

If this is the case and your pocket is somewhat shallow, it is worth looking out for secondhand items of industrial equipment. Most of this stuff is by no means worn out when scrapped. In fact it is probably being replaced purely because it is less convenient to use. Keep an eye on local adverts; if that bears no

Fig. 8/4 Electric drill with saw attachment.

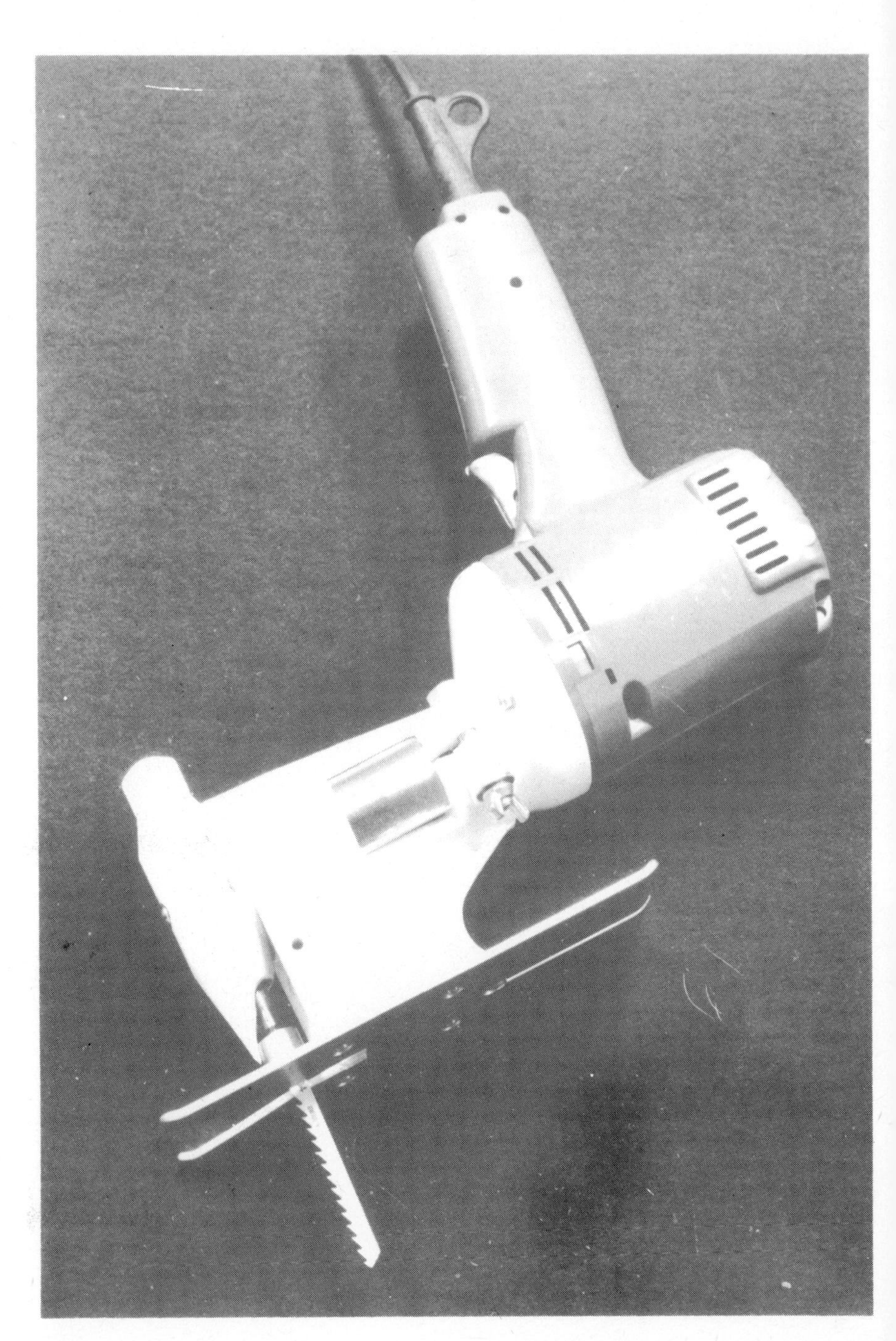

Fig. 8/5 **Jig-saw attachment, just the thing for cutting tortuous curves in wood.**

fruit, drop a line to local small or medium sized firms telling them what you are looking for, also offer to remove it, and, of course, enclose a stamped and addressed envelope for the hoped-for reply.

Motorising the workshop

As your workshop equipment grows you may find that you have acquired three or four small machines say, a small lathe, bench drill, grinder, circular saw, etc., and they all need to be powered. It is most efficient and convenient for each machine to have its own integral electric motor of correct power and speed rating for the machine tool it drives. However, there may be a number of reasons why we may prefer

Fig. 8/6 Simple amateur built lathe using an electric drill.

to drive them all from one source of power. This can be achieved either by setting up a lineshaft driven by the motor and with separate drives down to each machine tool. Or we may so plan our machine shop that our electric motor may be easily moved to each machine in turn.

Consider the lineshaft idea first. This is, of course, a move back to the original means by which factories were powered. A large steam, gas or oil engine, or even a water-wheel, drove a mainshaft which in turn drove lineshafts by large flat belts. The lineshaft then drove, usually via fast and loose pulleys, the counter-

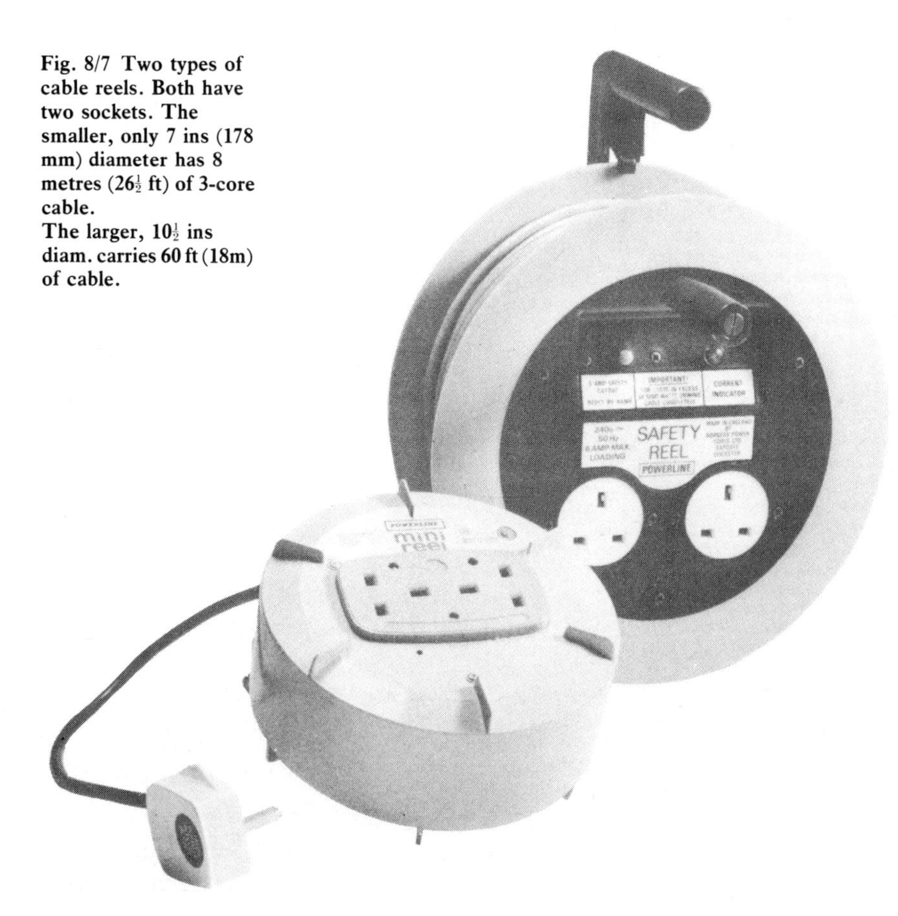

**Fig. 8/7 Two types of cable reels. Both have two sockets. The smaller, only 7 ins (178 mm) diameter has 8 metres (26½ ft) of 3-core cable.
The larger, 10½ ins diam. carries 60 ft (18m) of cable.**

Fig. 8/8 A geared type of drill chuck that may be used on drilling machine or lathe.

shaft of each individual machine. In a small workshop we can easily run a shaft completely along one wall driven by an electric motor; $\frac{1}{2}$ hp would be ample, with each machine belted to the shaft. We can do without countershafts, as only one machine will be in use at a time; unless we want to have the shaft running constantly, perhaps because we are emulating full-size practice by having our workshop steam- or gas-engine powered.

The lineshaft need be no more than 1 inch diameter, mounted in secondhand ball-races set in wood plummer blocks. One of the advantages of this system even today, is that it permits long belt drives, which work effectively with low belt tensions. We can use vee- or flat-belts as desired. The former somewhat more efficient and not requiring such accurate alignment; the latter permitting the fitting of fast- and loose-pulleys. The slackness of the long belt drives makes belt removal easy, hence no need for fast and loose pulleys or clutches, although the latter may be simply made if we want to do things properly.

The use of a simple movable motor entails no more work than setting up a lineshaft. It is a convenient method if the tools are of a type which run quite fast, e.g. all woodworking machines, tool grinders, drilling machines. Where a speed reduction is necessary e.g. for a metalworking lathe, the lineshaft may be more suitable, unless you build a small integral countershaft into the rear of the lathe.

For this arrangement, all the machines need to be mounted along a bench top, although if the workshop is wide enough, two opposite sides of a bench, standing in the middle of the workshop may be used. A reversing switch or crossed belts may be needed for the machines on the other side.

A round bar or pipe about 1 inch diameter and a flat bar are run along the back of the bench mounted as shown. The electric motor is bolted on a base of the type shown. There are a number of ways by which this can be made up from available material; even wood will do.

To move the motor position, slacken the clamping bolt, unship the belt, slide the motor along the bars,

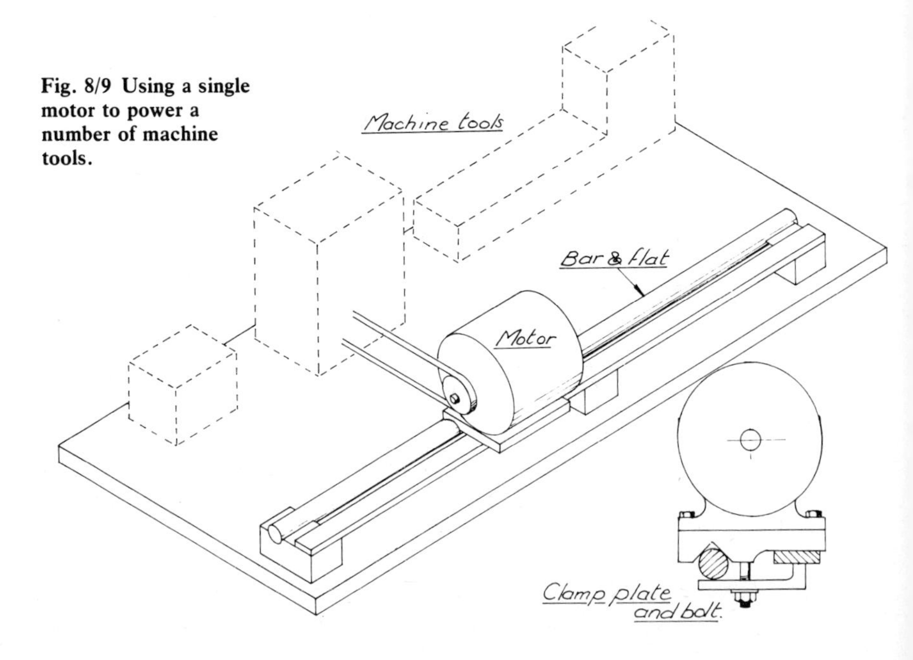

Fig. 8/9 Using a single motor to power a number of machine tools.

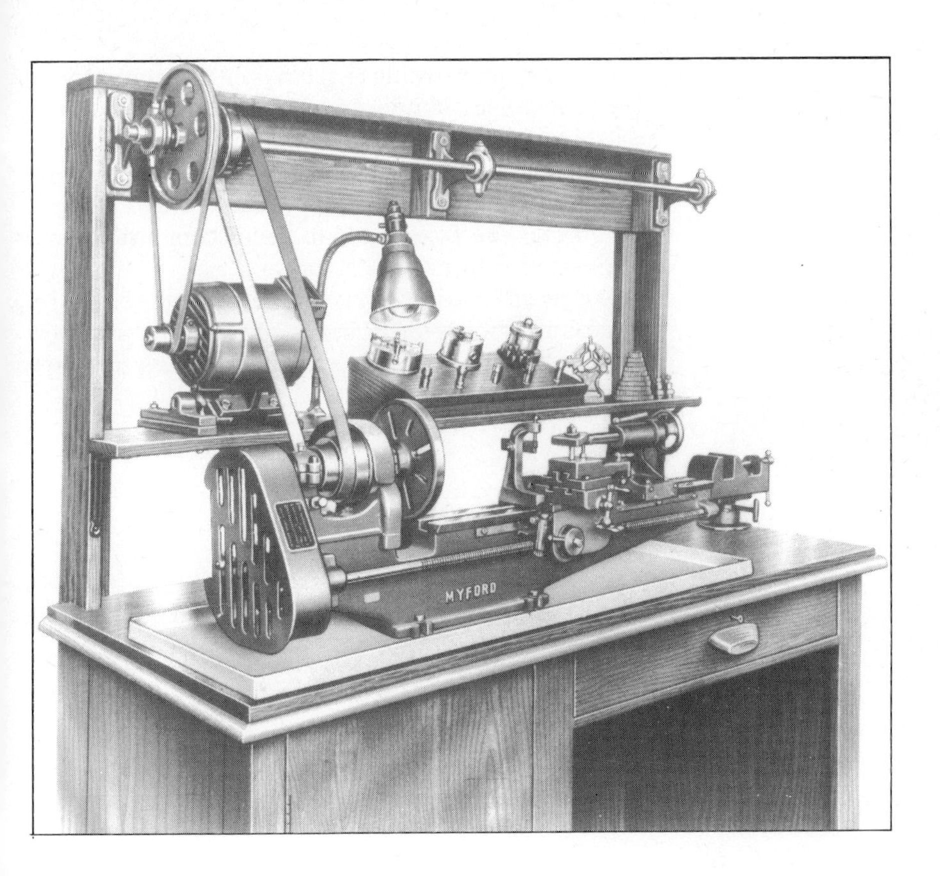

Fig. 8/10 A stout wooden bench fitted up as a machine shop with engineer's lathe and vice.

mount the next machine's belt, re-clamp motor and switch on.

In 1945 I bought for 10/- a secondhand $\frac{1}{4}$ hp electric motor and set it up as described to drive a small ASP (all spare parts) lathe, tool grinder, 6 inch circular saw, and $\frac{1}{2}$ inch "hand powered" bench drill. In 1950 all this was dismantled and the motor went on a 12 inch lawn mower. In 1960 the worn out mower was scrapped and the motor went on my $3\frac{1}{2}$ inch Portass lathe. It is still driving that after all these years – 10/- well spent.

Child safety

Finally to those readers with a young family. You must make the machinery in your workshop childproof. It is not possible to guarantee that the shed or room will be kept locked at all times. My own machines are and have always been fitted with three-pin plugs so that when unplugged they were electrically safe. The sockets were mounted as high up on the wall as possible to keep them away from inquisitive fingers.

When the children grew up, new sockets began to appear at a more reasonable height, but the possibility of future grandchildren makes me think that this was perhaps a mistake!

9. DO I NEED A LATHE?

It all depends on your particular interest. There is no reason at all why you should not have an efficient and interesting workshop, fulfilling all your creative needs and never have any desire to possess a lathe. On the other hand, there are those who can only be described as lathe enthusiasts and who spend their workshop hours solely devoted to producing work on a lathe. Such individuals often find ornamental turning in hardwoods, bone, ivory and brass a fascinating hobby and may be members of the Society of Ornamental Turners.

Purpose of a lathe

What does a lathe do? Well, very simply, it causes the work to rotate so that a cutting tool, which may be held in the hands when wood turning, or in a movable carriage in metal turning, can be applied to the rotating work. Various cylindrical forms may thus be produced.

In wood turning these will probably be decorative, such as lamp stands, chair and table legs, etc., while in the metals the turned work will be less artistic but more concerned with precision such as is necessary for spindles, pistons, wheels, etc.

In the amateur's workshop the so-called "lathe" may well be a multi-purpose tool consisting of a "headstock" carrying a rotating spindle. This spindle may carry a grinding wheel for tool sharpening and alternatively a circular saw blade for cutting wood. The addition of a "tailstock", moun-

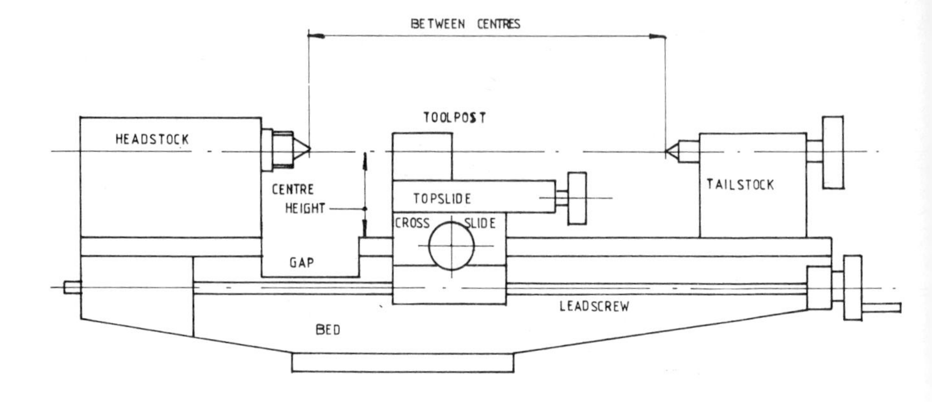

Fig. 9/1 This diagram illustrates the terms used to describe the various parts of a metal-turning lathe.

ted on and linked to the headstock by the "bed" turns the machine into a complete lathe, plus the fact that the tailstock permits drilling to be carried out, the lathe being virtually a horizontally mounted drilling machine.

Hence it is quite easy for a keen workshopper to rig up a spindle in a couple of bearings driven by an electric motor perhaps from a discarded washing machine. Add various bits and pieces to assist him in producing his numerous projects and he suddenly realises that he has the makings of a simple lathe, suitable for woodturning, and perhaps even work in plastics and the softer metals like aluminium and brass. This opens up an enormous new field of creative work including the fascinating craft of metal spinning.

It is this building-up system that forms the basis of the many ranges of accessories available for portable electric drills. The drill itself on a stand forms the powered headstock while the eventual addition of bed and tailstock completes a small lathe. The idea was shown in the previous chapter, Fig 8.6.

Specialist lathes

For the model engineering enthusiast, a more specialised machine is necessary. A tool-carrying carriage or "slide-rest" is a must in order to support the cutting tool against the greater forces encountered in machining metals including steel, and to permit the finer increments of tool movement necessary to achieve the required degree of accuracy of size and shape in the work. In addition, much slower speeds are necessary for metal turning, which means the use of a countershaft between the driving motor and the headstock spindle. For the very slow speeds needed when machining cast iron, a further reduction, usually via a "back-gear" is a necessity.

In the engineering industry lathes may be even more specialised and single operation "lathes" are common where high production rates are required.

Lathes suitable for the amateur who has visions of progressing to skill in turned work are not cheap and with little personal experience to guide him the newcomer may find it difficult to decide which type of lathe will suit him.

The lathes on the market fall into three groups, first, wood turning lathes, such as the Myford ML8, the Coronet, the Arundel and the Stokes. Most of

Fig. 9/2 A small lathe opens up a vast new field of creative work to the amateur workshop enthusiast – as experienced here by the author.

Fig. 9/3 Early treadle lathe, progenitor of many modern home workshop machines.

Fig. 9/4 The amateur's lathe has to be a versatile machine tool. Facing the port face of a large steam cylinder by fly-cutting on a small lathe.

Fig. 9/5 The EMCO "Unimat" 3.

Fig. 9/6 The "Perfecto" Lathe.

Fig. 9/7 The EMCO "Compact" 8 Lathes.

Fig. 9/8 The Myford ML7 Lathe.

them permit the addition of attachments for sawing, planing and other woodworking processes. If working in wood is your interest, then one of the above machines will be ideal.

The second group are the metalturning lathes consisting of the Boxford, the Myfords ML7 and ML10, Perfecto. These machines have all the normal screw-cutting and self-feeding facilities, and again, if light engineering is your interest you cannot, with due regard to the depth of your pocket, do better than choose one of these.

The third range are the very small or micro-lathes such as the Unimat, and the Cowell. These have many of the usual facilities of the lathes in group two but their capacity is more limited. However if your interests lie in 00 gauge model railways or tiny bits and pieces for model aircraft, such machine tools will suit you. (See the comments made in Chapter 15.)

A lathe for the amateur

Fig. 9/9 Earlier lathes of this type are still around and after reconditioning will give years of excellent service.

Now however we come to the real problem. If you just want to "have a go" at lathework, starting perhaps with simple wood turning and progressing eventually, you hope, to simple metal turning and model engineering, how should you proceed? You are probably loath to spend a lot of money on a

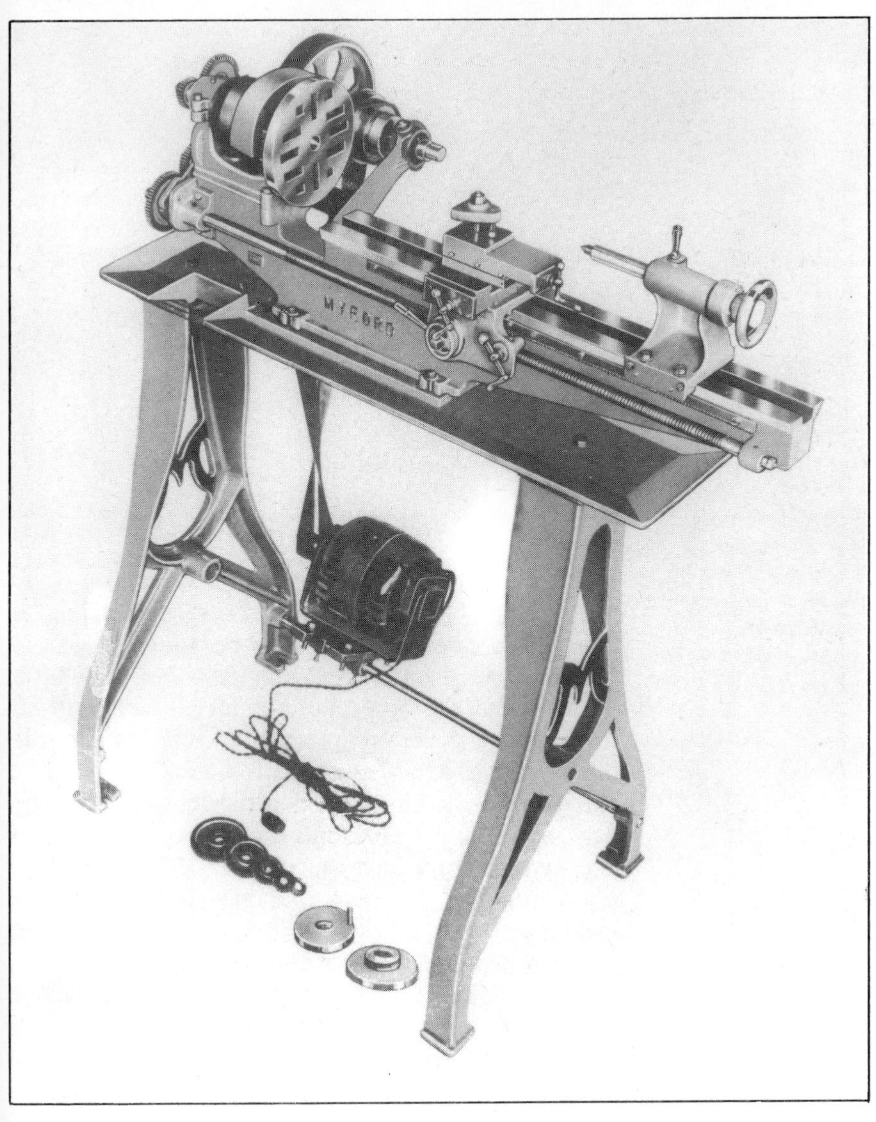

Fig. 9/10 The Myford ML10, a slightly smaller version of the '7'.

machine with facilities, the use of which you don't understand. On the other hand the micro-lathes of group three may just not seem big enough –?

My personal opinion is that a beginner is better with a SIMPLE lathe than a SMALL lathe and for this reason I make no apologies for using a very old illustration of what I consider to be the type of lathe suitable for a beginner. I don't demand that we go back to treadling, although it is an excellent way of getting to appreciate accurate tool sharpening. Unfortunately, as far as I know, there is no such machine on the market at the present day, but if there were I am sure it would be very popular.

Let us consider its merits. Bought as a simple headstock, tailstock and bed, it would serve for woodturning, etc. The addition of a slide-rest would permit simple metal turning followed by a countershaft and speed-reduction device to permit the machining of the driving wheels for that 3½ inch gauge locomotive which you now feel sufficiently confident to start. Perhaps a manufacturer may share my confidence that an up-to-date version of a lathe of this type would be worth marketing.

Fig. 9/11 A sophisticated machine for the home workshopper interested in advanced model or precision engineering. The Myford Super '7'.

10. EQUIPMENT TO CONSIDER FOR THE FUTURE

As our interest and pleasure in making things in a workshop increases, so will we be drawn towards some particular aspect of the work. We may find that the craft of hand and eye gives us the greatest pleasure, or that we find enjoyment in developing skill in the use of machines. A great friend who delights in forgework, will never saw a piece of metal if it can be cut on the anvil, or drill a hole if it can be punched or drifted. "After all", he says, "if I drift a hole I am not removing metal and weakening the bar as would drilling." For myself, my steam engine building interests demand the use of machines and I enjoy the pleasure of working at lathe and drill.

Machine tool skill in wood-working

As mentioned in the previous chapter, woodworking machinery for amateurs is often of the multi-purpose type, the lathe forming the basis. However, timber sawing is so frequently carried out that a separate circular saw may be felt to be an advantage. Sawing timber, especially ripping, takes a lot of space; hence, if the machine is made portable so that it can be used outside, it may be an advantage, if we have only a small shop.

The circular saw may be of the permanent type and with the availability of blades of many diameters and the necessary spindles, washers, bearings and pulleys from stockists, it is not a difficult job to make

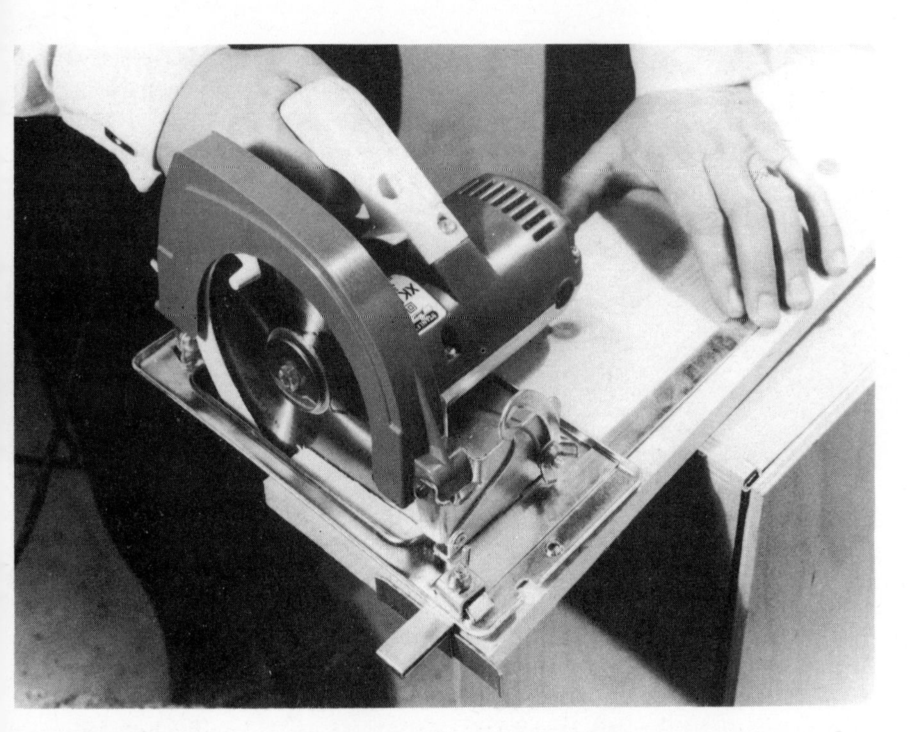

Fig. 10/1 An adaptable type of circular saw unit, with a powerful motor.

up a suitable frame and table in wood or steel.

An alternative which has great adaptability is the type of portable circular saw shown in Fig. 10.1. This may appear similar to that illustrated in Chapter 8; it is not however an attachment, but a unit in its own right and therefore usually larger and more powerful. Tables similar to that in Fig 10.2 are often available to permit the use of such a saw in the more traditional mode.

A band-saw is an extremely versatile tool for the woodworker which does not take up too much space. Until its introduction into the home workshop, the amateur will almost invariably think of his workshop projects in terms of straight lines. The bandsaw alters this, as curves and radii become as easy to produce as straight lines. This may be of particular interest to the person using his workshop in a part-time business capacity making wooden toys or similar items.

The very popular Burgess Bandsaw has been around for many years, and there has been a long line of models, each improving on the last.

The latest two-speed version gives even greater versatility and capabilities in the faster, neater cutting of a range of materials including of course, wood, but in addition aluminium, mild steel and plastics. It has a throat depth of 12 inches.

This new model has a low speed for cutting metal and a fast speed for wood; to alter the machine from one cutting speed to the other takes only five minutes and simply involves fitting a different wheel and drive belt.

Fig. 10/2 A saw table can convert the portable unit into the traditional type of permanent circular saw, so useful in the workshop.

Fig. 10/3 The two-speed bandsaw from Burgess Power Tools showing, in front, the additional wheel and belt supplied with the machine.

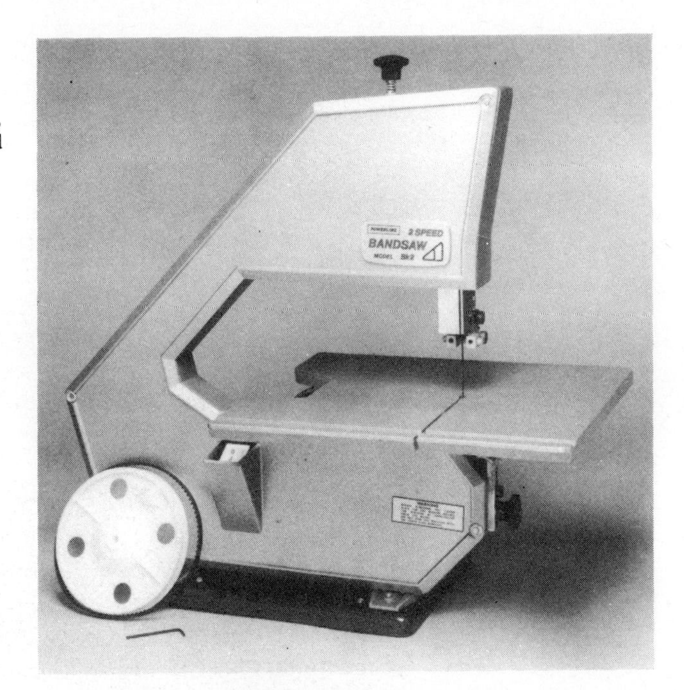

Fig. 10/4 The saw can cut to small radii through fairly thick timber. Table can be tilted and can carry a guide for straight cuts.

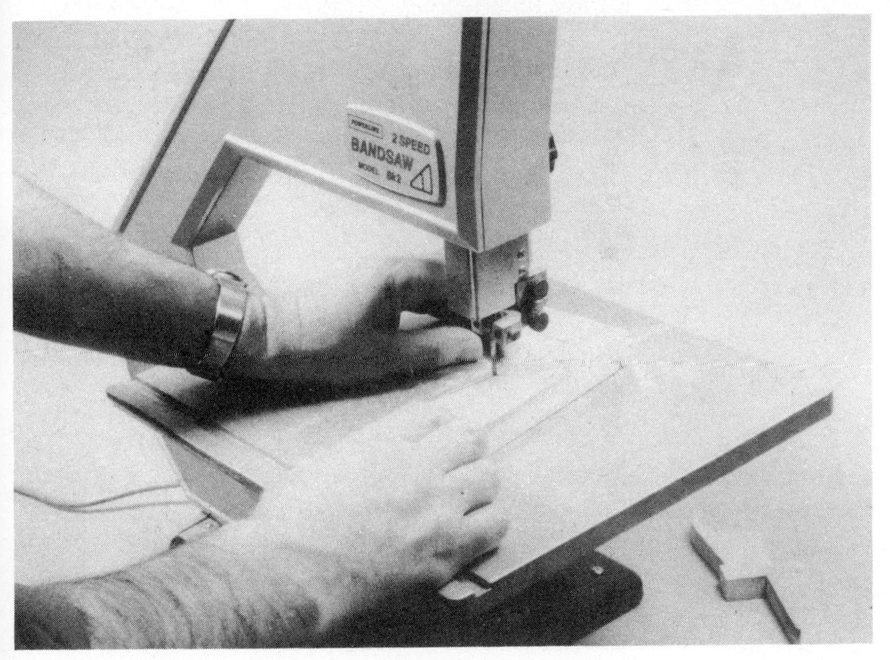

Fig. 10/5 An old friend, the first machine tool I ever had in my workshop.

By using this method instead of providing a two speed motor, the cost of the new model has been kept as low as possible.

For those many users who already have an earlier Burgess model a conversion kit to two-speed operation is available and again the actual conversion takes only a few minutes. The conversion kit provides three running wheels fitted with bearings, two belts and a motor pulley, plus full instructions.

A new feature of both the re-designed bandsaw and the conversion kit is the introduction of a sanding/linishing belt facility which is of particular value for sanding straight or irregular shapes on wood or metal. A medium grade belt is supplied as standard and two other grades are available.

Machine tools for amateur engineering

Without some device to rotate and allow us to apply endwise pressure, a twist drill is a pretty useless tool. This device, during the initial period of our workshop development, will probably be a hand brace or electric pistol drill as illustrated elsewhere in these pages.

While both of these are ideal for general workshop use, it becomes a little difficult to control and accurately position the drill for precise work, especially in metal, and some sort of drilling machine becomes the first tool sought by the amateur engineer.

The obvious choice for an electric drill is the type of stand referred to in Chapter 8. Although in my own workshop I have two drilling machines, capable of taking from very small drills up to 1½ inches diameter, I also have stands for my electric pistol drills. It is so convenient to be able to take drill and stand outside if drilling bars of metal too long to accommodate in the workshop. Two examples that come immediately to mind were, drilling black steel bar for a 7¼ inch gauge garden railway track, and building a boat (steam, needless to say) trailer.

At one time, machine versions of the simple hand drill were available, and Fig 10.5 shows the original machine of this type that was the first machine tool in

Fig. 10/6 The modern model engineer's lathe is a machine of great precision and adaptability.

my workshop many years ago. As I had no power supply, this machine was a godsend and I suggest that any reader in a similar predicament should look out for such a machine in second-hand tool shops, scrap yards, etc. A heavy flywheel makes operation smoother and an automatic feed mechanism has advantages.

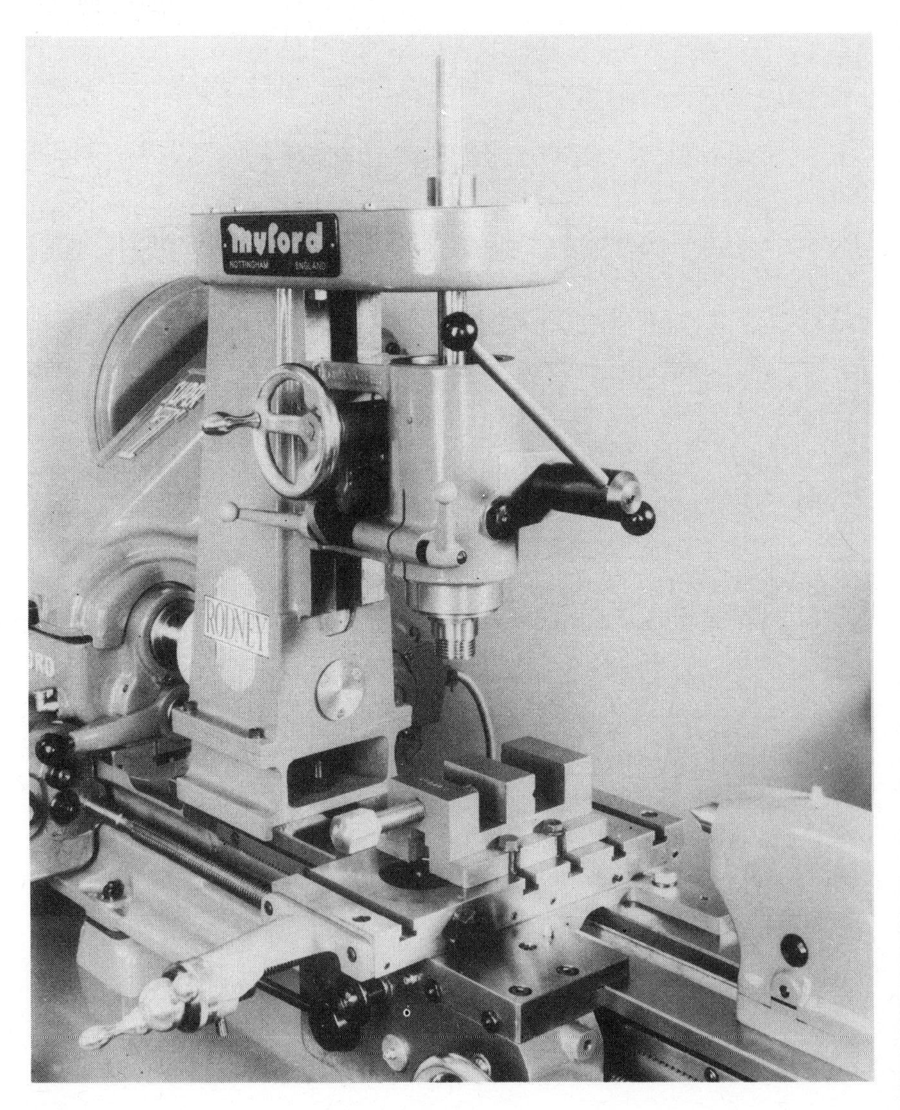

Fig. 10/7 For the advanced model engineer or amateur experimenter whose space is limited, a vertical milling machine, mounted as an attachment on a lathe has great potential.

Fig. 10/8 The PERFECTO Tool and Cutter Grinder.

For versatility in light engineering there is nothing to beat the small modern lathe as designed for the model engineer. Apart from normal turning operations, precision attachments permit such specialised work as gear cutting to be accurately performed. The development of the vertical milling machine as a lathe accessory opens up another new field of machining capabilities in the home workshop.

However, this is not without its disadvantages as the successful employment of these machines depends on the skill and facilities to regrind the, sometimes complicated, multi-edged cutters that are used. For this reason, many advanced amateurs are finding that a small version of the industrial type of tool and cutter grinder is a necessary piece of workshop equipment.

Where simpler machine tools are sought the shaping machine has the advantage of only needing simple tools, similar to those used on lathes. With ingenuity this machine has great possibilities and it is a pity that it is not more widely appreciated.

Making machine tools

There is no reason why the reasonably competent amateur should not build some of his own machines. Circular saw and bandsaw blades can be bought in any good tool shop, and designs for building such machines have appeared, from time to time, in MODEL ENGINEER and WOODWORKER. These pages show two views of the first tests being carried out prior to adding the finishing touches to a bandsaw built by the author.

For the metalworker, a powered hacksaw using ordinary hacksaw blades is not difficult to design and construct. A photograph shows one built by a late friend which contained all manner of devices for the intricate cutting of metal; in fact it almost took the place of a milling machine!

In a simpler vein, the linisher shown in Fig 10.11 is easy to make from bright steel sections and uses a commercially available abrasive belt. It is ideal for putting a lovely finish on the bright steel and brass work of engineering models.

Fig. 10/9 The PERFECTO 5 inch stroke hand operated shaper with automatic feed. The shaper is an adaptable machine, tooling is simple and, like the lathe, various devices can be employed to enhance its versatility.

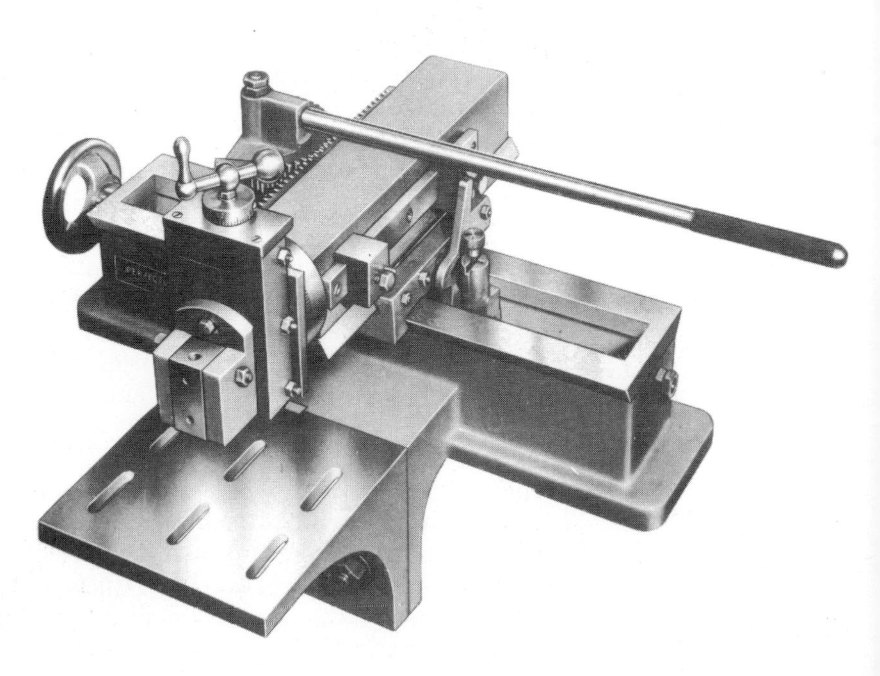

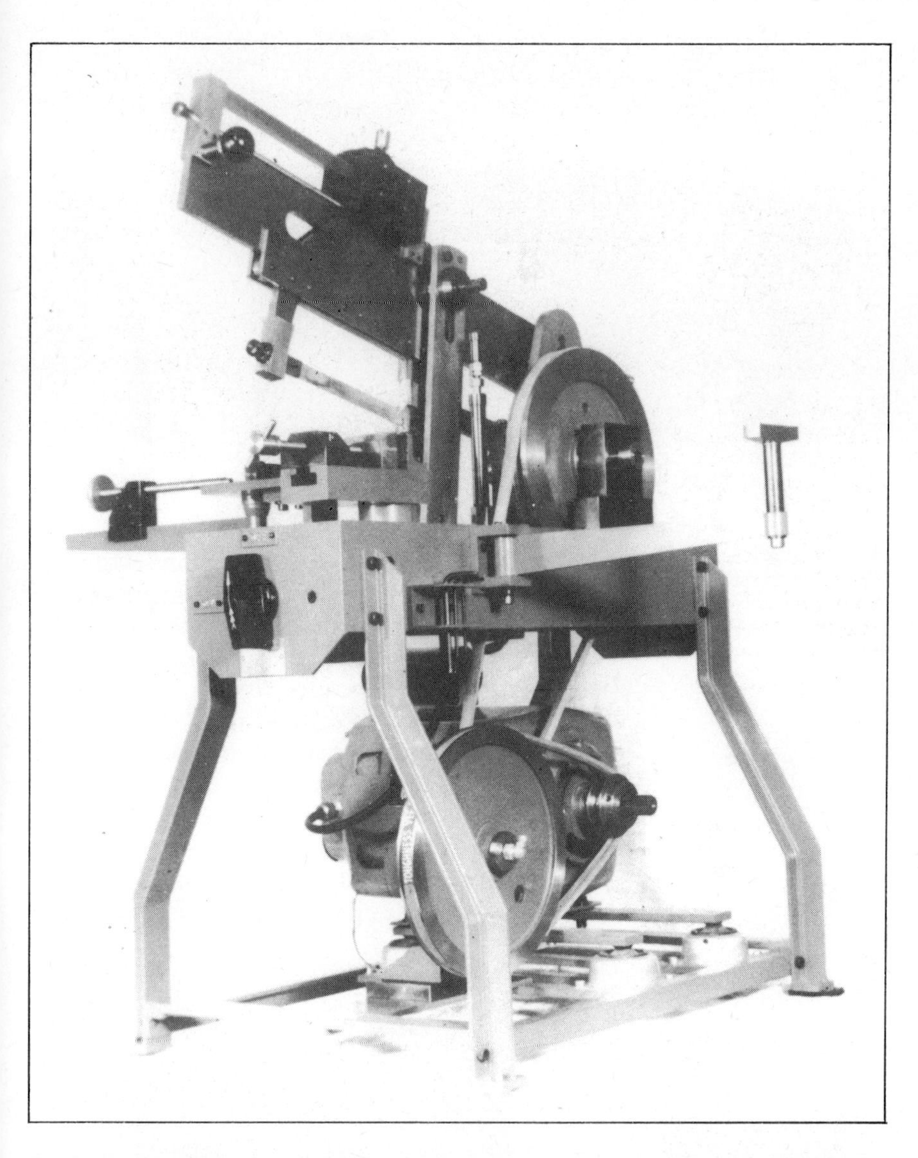

Fig. 10/10 A beautifully finished power hacksaw made completely without the use of castings. Based on an early MODEL ENGINEER design.

Another photograph shows the beginnings of a model engineer's lathe of 3 inches centre height designed by the author. As will be appreciated, the headstock bearings are separate items for ease of manufacture.

Drilling machines are not difficult to build and are

Fig. 10/11 A belt linisher for finishing metal parts. An easy-to-make project for the home workshop.

probably the first need felt by the home workshop enthusiast. Where precision requirements are not too demanding much of them can be made up from "off the shelf" bearings, shafts, pulleys, etc., obtainable from various advertisers.

Some years ago I was required to produce a design for, and manufacture, a small batch of bench drilling machines which had to be simple and cheap. The illustrations Fig 10.13 (a & b) show one of the resulting machines. As can be seen in the photographs, the heads were not adjustable on the columns. After all, how many people climb on the bench to raise or

Fig. 10/12 The beginnings of a small 3 inch model engineer's lathe.

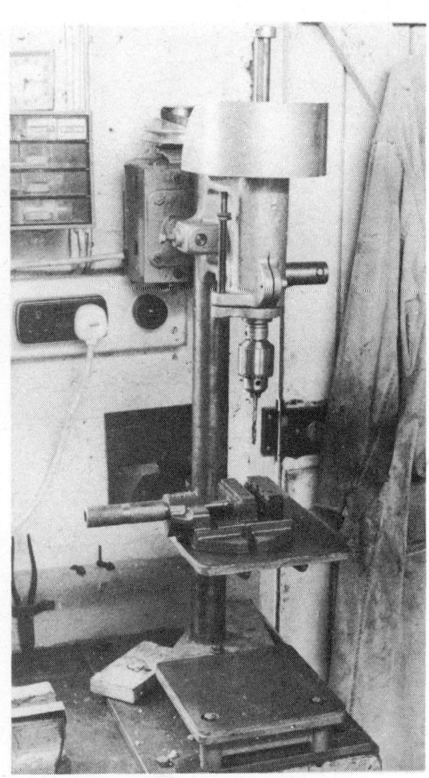

Fig. 10/3 (a & b) **Two views of a useful amateur-built $\frac{1}{2}$ in. capacity bench drilling machine.**

lower a drill head? The quills were cut from hydraulic cylinder tube so no machining was needed apart from facing to length and cutting the rack. The machines tended to be designed around what could be bought, but although built down to a price, they gave good service.

If a lathe is available, there is no great difficulty in building your own drilling machine, especially if you employ a simple sensitive feed rather than the rack and pinion system normally used. An example of such a scheme is the "M.E." Drilling Machine designed many years ago by Edgar Westbury. Drawings are available from the M.A.P. Plans Service (Plan W.E.2.) and castings from Woking Precision Models.

For the modeller concerned with small work, the

Fig. 10/14 (a & b) **Two views of an amateur-made bandsaw undergoing its initial tests.**

miniature drilling machine (Chap. 15), by Cowells Engineering Ltd., of Norwich is to be recommended. It is built to a high degree of accuracy and matches their small model engineering lathe, with which it shares some parts. The same company make a micro-size machine of a specialised nature, for drilling printed circuit boards and other very fine work. Unlike the normal drilling machine, this version does not use a chuck to hold the drills; instead, they are held in pulleys and run in vee blocks at some 15,000 rpm. It has a capacity of from 2 mm (0.080 in. approx) down to 0.35 mm (0.013 in.); a bit small for

most of us "model mechanics". It is interesting however to study the alternative design principles employed and to consider whether they might lend themselves to the development of a simple type of drilling machine for our own workshops.

Comments have been made earlier of the advantages to the "home workshopper" of using the classes run by local evening institutes and technical colleges as a useful adjunct to his own meagre equipment. The building of machine tools of various types is a perfect example of the projects that can well employ a session spent at such an establishment.

11. WORKSHOP ELECTRICAL INSTALLATION

contributed by E. Woodward, Lecturer in Electrical Installation.

The garden shed type of workshop provides a very suitable workplace for those many projects that you have in mind but I would suggest that you consider what is required in the way of electrical supplies to be installed in the workshop.

Stroboscopic effect

Adequate lighting is an essential requirement for any workshop, in order to provide a well lit area for safe and efficient working. The fluorescent gas discharge lamp unit is particularly suitable for the provision of general lighting in your workshop, giving a good spread of light with minimum shadow effect.

This lamp produces a much greater light output than the same power-rated ordinary filament lamp.

This high efficiency light source will therefore give much better results than the filament lamp and run at a much lower lamp temperature.

The fluorescent lamp produces its light output when an electric current is passed through the gas present in the lamp, and if the lamp is connected to the mains alternating current supply, the current flowing through the lamp will produce flashes of light at one hundred times each second.

If the fluorescent lamp is to be used for direct lighting it may be necessary to consider the likelihood of stroboscopic effect being created by the flashes of light shining onto rotating machinery.

Lighting and tools

A striking example of stroboscopic effect is often seen on television or in the cinema when the wheels and spokes of the stagecoach appear at different instances, to be turning forward at a slower pace than they should be, remaining stationary, or revolving backwards. This optical illusion is created by the shutter speed of the camera taking the film, pulsing at a speed related to the wheels and spokes of the stagecoach.

The same effect can be produced by the fluorescent lamp when used to illuminate a rotating object.

Consider a four jaw lathe chuck rotating at 1500 revolutions per minute and illuminated by a fluorescent light source producing flashes of light at 100 times per second (6000 times per minute). The light will produce four pulses of light in one revolution of the lathe chuck, therefore, each jaw of the chuck will be lit each time it turns through 90°. Under these conditions it is a possibility that anyone looking at the chuck could mistakenly believe that the chuck is stationary.

If the chuck is of the three jaw type and revolving at 2000 r.p.m. the same effect may be produced.

It is very unlikely of course that these exact speed relationships will occur, but should you wish to safeguard against the danger of this effect being produced, you can instal a twin lamp antistroboscopic unit.

This lamp unit is designed to eliminate the stroboscopic effect by using two lamps in the same unit producing light outputs out of phase with each other.

Power tools

You will no doubt wish to use portable tools in your workshop and provision of two sockets should be adequate for this need; with regard to comfort heating, the second socket can be used for plugging in a convector heater.

Many types of portable tools are on the market and when you buy them, the double-insulated type is

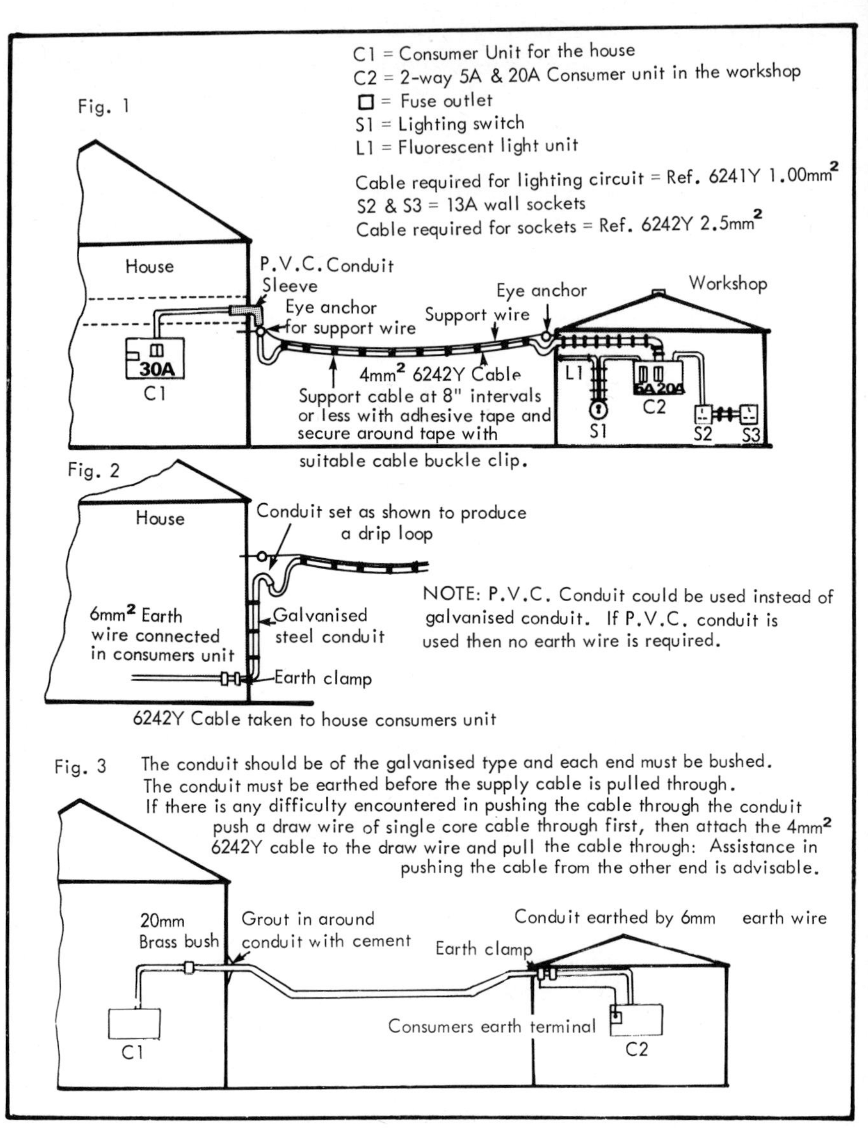

Fig. 11/1

recommended. Double-insulated portable tools are designed with the body, switch, handle and side handles made of a material that will not conduct electricity. This design ensures that these tools are extremely safe to use. Such tools can be easily recognised by the symbols shown, which are printed on the body of the tools. You are thus reassured.

Portable tools other than those mentioned must be effectively earthed and it is advisable to have this type tested at regular intervals, to ensure that the earth wire is securely attached and providing a continuous circuit from the earth pin on the plug top to the body of the tool.

Installing electrical supply

If you feel capable of carrying out the electrical installation for the workshop, it is important to appreciate that it should be installed to meet the requirements of the current edition of the "Institution of Electrical Engineers' Regulations for the Electrical Equipment of Buildings". The essential requirements to be met are;

(a) to choose the correct type and size of cable,

(b) to provide adequate protection for the cables in the form of fuses or circuit breakers, these latter being devices which will open the electrical circuit if an excess of current flows in the circuit,

(c) to fix cables securely and provide adequate protection for these cables against mechanical damage,

(d) to make good electrical connections where conductors are terminated,

(e) to provide adequate earthing of all metal associated with electrical equipment,

(f) to test and inspect the electrical installation before connecting it to the supply mains.

This last requirement would normally be carried out by a recognised electrical contractor.

Note! Before an additional installation is to be connected it is essential to check if the existing installation to which it will be connected can withstand the additional load. If in doubt consult your local electricity supply authority.

To supply the workshop with electricity you will need to take a suitable size cable from the house supply consumer's unit and if a spare 30 amp fuse outlet is available in this unit, then no problem arises.

However if there is no spare fuse outlet, it may be necessary to purchase an extension unit and fit it alongside the existing consumer's unit in the house.

Another possible method of overcoming this problem is, if you have two lighting circuits rated at 5 amp each and the total lighting load on the two circuits does not actually exceed 5 amp, you can connect these two circuits on to one 5 amp fuse and convert the spare 5 amp fuse outlet into a 30 amp outlet by changing the fuse carrier and base to a 30 amp type. The required 30 amp fuse carrier and base may be obtained from an electrical retailer.

The fuse element to be fitted in the fuse carrier must not be larger than 25 amp as this is the nearest fuse size to suit the cable supplying the workshop, which should be Ref No 62427Y – 4 mm^2.

The next job you will need to consider is how you intend to run the cable to the workshop and much will depend upon the particular situation. You have three choices;

1. the catenary system,
2. the conduit system,
3. the mineral insulated copper sheathed system.

The catenary system consists of the supply cable from the house to the workshop being securely suspended on an overhead wire support spanning the space between the two buildings. A suitable support can be a pvc covered clothes line. Don't use bare wire for this purpose, because, as you will appreciate, the metal support wire will then need to be earthed for safety purposes.

Fig 11.1 shows the method of installation where it is convenient to pull the cable under the floor at first floor level. You will need to drill or chisel through the wall to enter the cable at the point shown; the size of the hole being large enough to allow a pvc or steel conduit of 20 mm or above to be inserted for cable protection.

If it is required to enter the cable at ground floor level then run the cable through the length of steel conduit (20 mm diameter) from the catenary position, as shown in Fig 2 of Fig 11.1.

The conduit system as shown in Fig 3 of Fig 11.1, consists of one length of 20 mm galvanised steel conduit, bushed at both ends, erected between the two buildings to provide an enclosure for the cable. This system may be used where a substantial support exists between the two buildings. The conduit can be set to the required shape by using a "setting block", this consists of a suitable clearance hole drilled in a hardwood block about 3 feet long by 6 inches wide and 2 inches thick; individual lengths of conduit may be joined using conduit couplers. Note that any exposed threads should be painted to avoid corrosion.

The conduit required will be seam welded, galvanised, screw thread type and it should be securely fixed to walls, etc., by galvanised steel saddles.

The third choice is the mineral insulated copper sheathed system and this is mainly used where the cable needs to be laid underground. Steel conduit is not a suitable material to be buried in the ground because of the likelihood of corrosion due to dampness and acidity present in the soil. The MICS system uses a conductor which, because it is insulated with a mineral which absorbs moisture, makes it necessary to effectively seal the cable at both ends to complete the installation. The operation of sealing the cable requires special tools and a reasonable degree of skill. Therefore, I suggest that the "sealing off" of the cable ends be left to a recognised contractor who will supply the cable to the required length, with ends sealed, ready for laying.

The type of cable you will require will be Light Duty CCV2 L2.5 and the seals will be of the Wedge Pot Type with Earth Tail.

Now to the task of installing. A trench with a depth of at least 18 inches will be necessary and the bottom of the trench should be covered with sand. When this has been completed, the cable is placed in the trench and then covered with bricks for mechanical protection. The trench can now be filled in.

The extremities of the cable left exposed at each end can then be fixed to the structure by suitable clips. It is not necessary to run this cable through the house or workshop right up to the consumer's unit; instead the ends of the cable can be terminated in a 30 amp 3-way plastic joint box fixed at a suitably protected position. The cable from this joint box to the consumer's unit will then be 4 mm^2 6242Y, which is a twin cable with earth conductor rated to carry 27 amps.

Inspection and testing

After the installation has been completed it will be necessary to inspect and test it. Inspection must ensure that:

1. Cables are fixed securely and are not under undue strain.
2. Accessories are securely fixed and conductor connections electrically sound.
3. Conductors are connected to the correct terminations.
4. Earth continuity conductors are sleeved with green sleeving at accessory points.
5. Entry holes in walls are made good with cement.

To test electrically that the complete installation meets the necessary requirements, the test would be performed at twice the voltage than will eventually be connected to the circuit. For these tests you will need various instruments, an insulation tester, commonly called a "Megger" and a continuity tester to measure the earth continuity from all outlet points. Both instruments can be combined in one meter and a selector switch on the meter gives the choice for insulation testing to read Megohms (M Ω) or ohms (Ω).

To check for earth continuity, one lead of the instrument is placed on the earth point at the outlet and the other lead placed at the main earth point in the consumer units. Completing the circuit should show a reading not exceeding 1 ohm; where steel

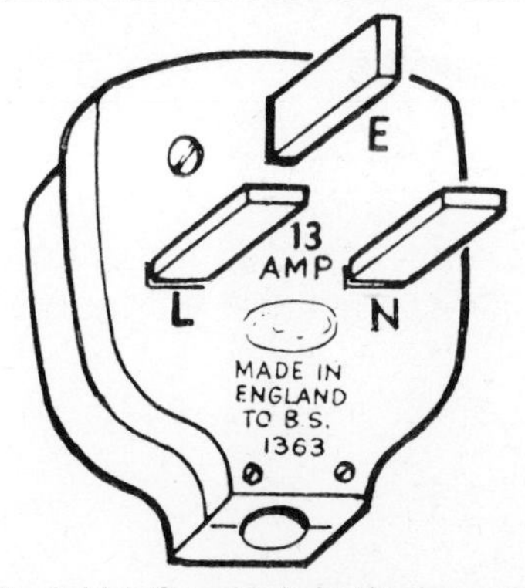

The British Standards Institution marking on a piece of electrical equipment. Quoting the number of the Standard that describes the details of the Standard.

The British Standards Institution Kite Mark.

The British Electrical Approvals Board certification mark for domestic electrical appliances.

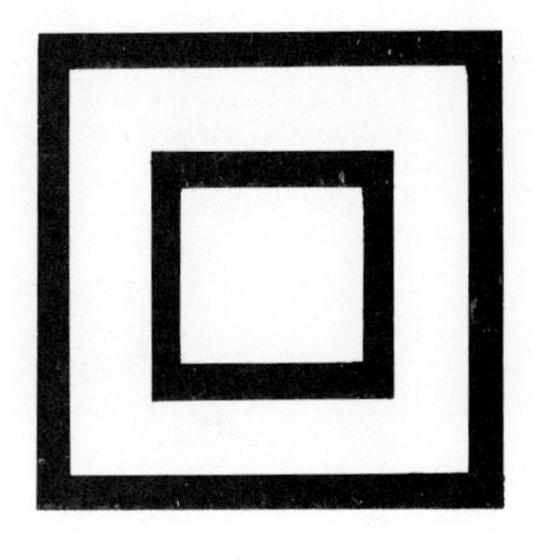

The Square in a Square Mark which signifies that the piece of equipment is designed and built as an electrically double-insulated appliance.

Standard of Quality of Materials and Construction for Electrical Equipment.

To ensure a high standard of manufacture and safety in the electrical apparatus you purchase, look for the various standards markings.

The statement that an item is made to B.S. (British Standard) followed by a number is one method of showing that the appliance complies with the necessary standard covered by that number.

Alternatively, the Kite Mark, made up of the two letters 'B' and 'S' may be displayed; or the British Electrical Approvals Board roundel, which carries, in its centre the B.S. Kite Mark, may be used.

Electrical appliances, especially portable power tools, which are 'double-insulated', and therefore extra safe, carry a 'square-within-a-square' sign.

When you buy electrical equipment look for these signs of safety and quality.

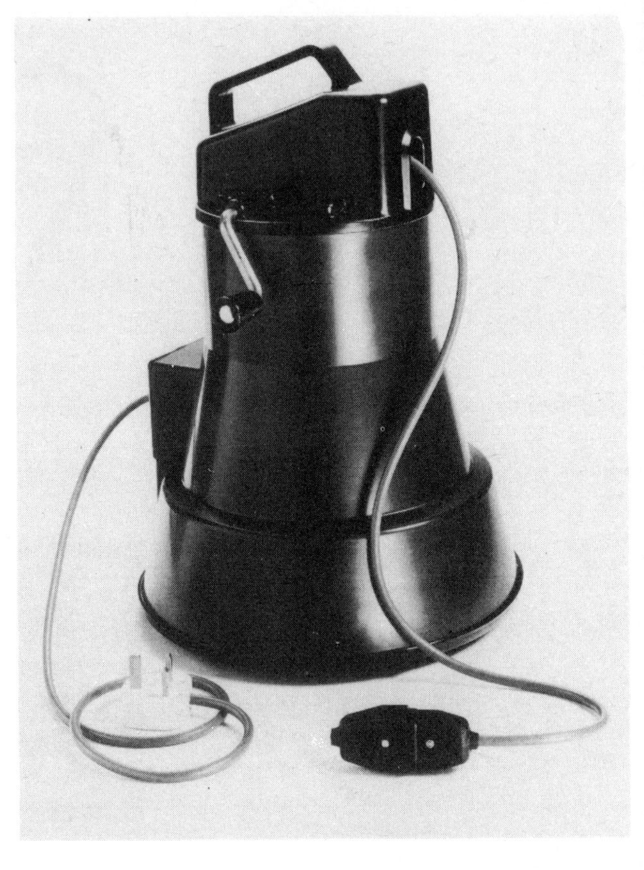

Fig. 11/3 A neat, safe way to keep extension cable under control. Has a geared drive for quick rewind, is double insulated and comes complete with 100 ft of 3-core cable, mains plug and safety cable connector. From STANLEY Power Tools.

conduit is used the reading should not exceed ½ ohm.

To check for good insulation throughout the installation it will be necessary to check that all switches are switched on. NOTE THAT NO CONNECTION IS MADE TO THE MAIN SUPPLY AT THIS STAGE.

Make sure that all fuses are in place, now place the end of each meter lead to the red and black conductors marked L and N respectively at the consumer unit in the workshop, switch on the meter, the acceptable reading is 1 Megohm or more. Now carry out the same test from L to earth E and next from N to E. The acceptable reading is 1 Megohm or above.

If you carry out the tests as stated and your

readings come within the acceptable values then you can switch on the supply by making off the cable at the house position.

Finally test the sockets to ensure that the live conductor is connected to the correct terminal. For this you can use a neon pocket screwdriver designed for 250 volts. The neon lamp should light if the right hand terminal, looking from the front of the socket, is connected to the screwdriver end of the test lamp. Any lighting switches must be tested in the same way. With the switch cover of the switch removed, the tester is connected to the outgoing lead to the light; with the switch off, there should be no light but with the switch on the neon lamp should light.

When you have completed your installation it should be trouble-free for many years but I do recommend that every five years a complete test be made to ensure that it is still up to standard.

If you are at all doubtful of your ability or knowledge, entrust the work and testing to an electrical contractor making sure that you employ a member of the National Inspection Council for Electrical Installation Contracting (NICEIC). This contractor will have all the necessary test equipment and is qualified to carry out this work.

12. THE LEGAL IMPLICATIONS

contributed by A.W. James, LL.B., Solicitor.

The previous chapters of this book have suggested a number of locations for your workshop – attic, room in the house, garage, garden shed or basement. Each has its particular problems. Here I will try to cover some of the different legal implications of the location you may choose – perhaps some of the points raised may even influence your choice. One cannot, of course, hope to cover any matter in great detail; all that can be done is to awaken you to some of the complications and try to help you to avoid them.

I will deal, in greater or lesser detail, with the following topics – planning permission, building regulations, listing of a building as of Special Architectural or Historic Interest, Conservation Areas, Tree Preservation Orders, restrictive covenants and rateable value.

Planning permission

Foremost in your mind will be the question "Do I need planning permission?" I think in answer to that question it would be of some benefit if I were to explain in general terms the circumstances in which planning permission is and is not required.

Basically, the Town and Country Planning Act 1971 provides that planning permission is needed for the carrying out of what is termed "development". This is defined as –

"(a) the carrying out of building, engineering, mining or other operations in, on, over or under land; or

(b) the making of any material change in the use of any building or land."

There is no comprehensive definition of "building operations" in the Act although it is defined to include "rebuilding operations, structural alterations of or additions to building and other operations normally undertaken by a person carrying on business as a builder." The definition is really wide enough to take in almost anything. "Engineering operations" is even less well defined, merely being stated as including "the formation or laying out of means of access to a highway." Hardly very helpful! "Mining operations" are unlikely to concern us but what of "other operations". What are they? Well, you are not alone in asking that.

You may find it difficult to think of something you would describe as "operations" or "works" which would not be building, engineering or mining. So do those who have the job of interpreting the provisions. Suffice it to say that it is a very restricted area unlikely to be of concern to you in the creation of a home workshop.

What about "material change of use! What is it? This is really two questions. Firstly, what is a change of use? and secondly, is that change of use "material" i.e. substantial?

A change of use is really a change in the purpose for which a building or land is used, but before we can decide whether there has been a change in the purpose we will have to discover what the "existing" use of the building or land is. In this respect we must look at what is called the "planning unit". In the case of a single occupancy house, bungalow or flat this will usually be the house and garden as a whole. The use of a house is clearly residential and the creation of a workshop in a house is unlikely to change the overall character of the residential use; it is purely incidental to that use. The Minister of Town and Country Planning in the past offered the following advice "the point at issue is whether the character of the whole existing use will be substantially affected by the change which is proposed in a part of the building."

Certain operations or uses are, however, specifically excluded from this definition by the Act. Briefly these are (to the extent that they are relevant to this book) –

(a) works for the maintenance, improvement or alteration of a building which affect only the inside of the building or do not materially affect the external appearance and which do not provide additional space in the building below ground;

(b) the use (but not the erection) of a building within the curtilage of a dwellinghouse for any purpose incidental to the enjoyment of the dwellinghouse as such.

Activities falling within either of these would not require planning permission because they do not constitute development. In addition certain other operations and uses which are development do not necessitate a specific application for planning permission since they already have the benefit of a general permission granted by the Minister in a General Development Order. Development of this kind is usually referred to as 'permitted development'. The relevant provisions specify the following –

(a) the enlargement of a house (including the erection of a garage) provided that the extension –

  (i) does not exceed the height of the original house;

  (ii) does not extend in front of the forwardmost part of the house;

  (iii) does not add more than 50 cubic metres to the house (as originally built or as it was on 1st July 1948 if built before that date) or 10% of the cubic area of the house, up to a maximum of 115 cubic metres, whichever is greater

and

(b) development, other than the erection of a house or garage, within the curtilage of a dwellinghouse for the enjoyment thereof provided that –

(i) no building shall extend in front of the forwardmost part of any wall of the house which faces a road;

(ii) no building shall exceed 3 metres in height (or 4 metres if it has a ridged roof);

(iii) the total area of the garden covered by buildings should not exceed 50%

Where then does putting all this together and applying it to the home workshop situation get us? Let us deal first with a workshop that does not involve any actual works, merely the use of an existing space for your workshop. As long as yours is intended to be solely a hobby workshop then the use of existing facilities will not need planning permission, but should it develop beyond that to the position where you are using it to produce goods to sell you will probably need permission for the change of use.

If works are involved, what then? I will take a look at each possible location in turn.

ATTIC – Are the works "development"? That is, do they fall outside the general category of "works for the maintenance, improvement or other alteration of a building not materially affecting the external appearance of the building"? If the answer is "No", then planning permission would not appear to be needed. If "Yes", then we have to consider a further question "Will the works increase the volume of the house by more than the limit allowed? If "No", we must dccide whether the allowance has already been used. If it has been then permission will be needed.

BASEMENT – Clearly the creation of an additional room below ground is "development" but there is nothing to stop you taking advantage of the category of permitted development which allows for a limited increase in the size of dwellinghouse.

EXISTING ROOM IN THE HOUSE – provided you do not intend to do anything like create a new window space that will materially alter the look of the outside of the house, there should be no need for you to submit a planning application.

GARDEN WORKSHOP – provided you have not already saturated your garden with other outbuildings, there should be no need for you to submit a planning application for any of the types of building described in this book.

GARAGE/WORKSHOP – Will it be an extension to the existing house or a separate building? If attached, can it come within the permitted limits for extension, otherwise it will necessitate the submission of a planning application. If detached, planning permission is likely to be required.

These notes are intended to be solely a guide and should not be taken to be a definitive statement of the law on any particular matter, since it is impossible in the space available in a book of this sort to cover all the permutations which could arise due to circumstances peculiar to a particular case.

In each case it is, of course, sensible to check whether planning permission has already been granted to a previous occupier. Even then you are not out of the woods because there are strict time limits within which development must be commenced or there may be conditions attached to it but at least a previous permission will indicate the way in which the planning authority have previously viewed the situation.

One thing you should know is that if you carry out works which need planning permission and don't obtain it you can be required to reinstate the property to its former condition. Hopefully, however, it won't come to that because there is a procedure whereby you can ask the planning authority to determine formally whether planning permission is considered to be needed. Even that will probably be unnecessary as I am sure the planning department of your local district or borough council will be willing to advise you informally as to whether you need planning permission for what you propose. They invariably try to be quite helpful.

Building regulations

Approval under the Building Regulations is usually dealt with at the same time as the application for planning permission, although they are, in fact, completely separate matters. Building Regulations govern construction standards, materials to be used, ventilation and sanitation standards etc., in the erection of a new building (which includes part of a building) or when a material change in the use of a building is proposed. The Regulations are extremely complex and detailed and virtually anything involving works will need building regulation approval. In particular if you are thinking of using the attic for your workshop bear in mind that the working area must have a floor-to-ceiling height of at least 7′ 7″.

What about the actual mechanics of the application? Well, if you propose to undertake works which come within the scope of the Building Regulations there is a duty on you to submit plans of those works to the local authority. As a general rule anything which needs planning permission will need building regulation approval.

Once the application is lodged with them, the local authority – again this will be the district or borough council for your area – has five weeks, or if you consent, two months, to determine the suitability of the plans. If the plans comply with the Building Regulations the local authority has no option but to approve them. Only if they do not can they reject them. Disputes may be determined through the Magistrates' Court or, in certain cases, by the Minister.

If any work is carried out in contravention of the regulations the local authority can require the owner to alter them or even pull them down. They do this by serving a notice on the owner. If the notice is not complied with, the local authority may pull down, remove or alter the works, charging the cost to the person on whom the notice was served i.e. the owner, provided the notice was served within one year of the completion of the works. This need never happen if the regulations are observed.

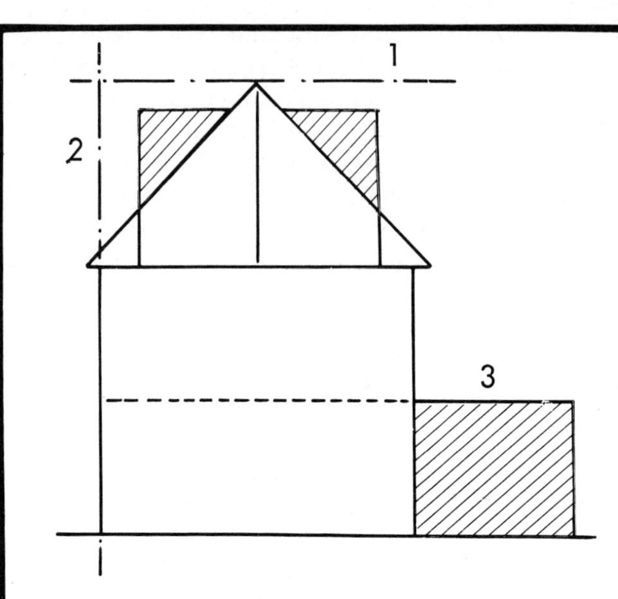

PLANNING PERMISSION

Unless your house or the locality is subject to a special order, you will not need planning permission, if:—

1. Your conversion does not exceed the height of the original house.
2. It does not come in front of the fowardmost part of the building.
3. It does not exceed, together with any other addition to the property, 50 cubic metres or 10% of the cubic area up to a maximum of 115 cubic metres.

The existing ridge height must be able to allow a finished floor-to-ceiling height of 2.3m (7′ 7″) to produce a habitable room.

Buildings of Special Architectural or Historic Interest and Conservation Areas

Over 200,000 buildings in Great Britain have been listed by the Secretary of State for Environment as being of "special architectural or historic interest." Copies of these lists are kept by both the local District and County Councils and are open to free inspection. The listing must also be registered as a local land charge and when a building is listed, notice of the fact must be given to the owner and occupier of the building.

There is no power to object to the listing of a building, except in the indirect sense that appeal does lie against refusal to allow works for its demolition, alteration or extension.

Once a building is listed anyone who carries out any works for the demolition, alteration or extension of that building without obtaining consent commits an offence, which is punishable by fine or imprisonment.

Similar provisions relate to buildings which although not "listed" buildings are in what are known as "Conservation Areas". These will be prescribed by your local councils following public consultation. Applications for what is usually termed "listed building consent" are made to your local district or borough council. If they turn down the application you can appeal to the Secretary of State.

A proposal to build something in the garden of your house could involve the removal of a tree or two or the loss of a number of other plants. Here again there are statutory provisions of which you could unwittingly fall foul. Firstly, you must give six weeks' notice to the local district or borough council if you intend to fell a tree which lies within a Conservation Area. Secondly, certain trees may be subject to a more stringent form of control known as a Tree Preservation Order. Strictly, if a tree is subject to such an order you need the consent of the authority which made the order, which could be either the district or county council, to do almost anything to the tree.

In addition the Conservation of Wild Creatures and Wild Plants Act 1975 prescribes a variety of plants the destruction of which is an offence. Fortunately the sort of plant involved is unlikely to be found in your garden, although it is not impossible.

Restrictive Covenants

There may be a restriction in the deeds of your property which limits the way in which you can use the house and garden. The sort of restriction one usually finds, however, is unlikely to prevent the creation of a home workshop, although there may be a restriction on the height or location of anything you propose to build. So if you have the deeds, check them; if not, because, for example, the property is subject to a mortgage, then the solicitor who dealt with the conveyancing when you purchased should be able to tell you or find out for you.

Rateable Value

The final issue of which you should be aware is the implication of what you propose upon the rateable value of your house and therefore on the amount of rates you actually pay.

The rateable value of a property is based on the figure which the District Valuation Office considers you could obtain if you were to let your house. Clearly therefore anything which increases the letting value of a property is going to increase its rateable value.

Certainly if you erect a garage/workshop the amount of your rates will increase. Similarly if you convert an attic or create a basement, but probably not so if you merely make use of an existing room, but take note that if you live in anything but a detached property you should think carefully about this as a location because a neighbour is not likely to take kindly if you regularly use noisy machinery.

Well, as I have said, I have not attempted to cover any matter in great detail. I hope however that I have given you some food for thought. I cannot stress enough however that there are usually people at your local council offices willing to offer help.

13. HOW TO GET BUILDING PERMISSION

Having decided you do need planning permission and presuming that the rest of the family tell you to go ahead and build that workshop in the garden, the next step is to obtain permission from the local authority.

Our writer on the legal implications of a workshop has defined the requirements as demanded in law, but just exactly how do we go about dealing with the necessary paperwork?

First, let us appreciate a simple fact which is, that when we build a workshop, or any other structure for that matter, in our garden, we are carrying out a DEVELOPMENT. And although we may feel that we have a perfect right to develop our own property as we please, we would hardly thank our neighbour if he developed his garden by erecting an enormous block of flats completely blocking out sun and view from our little semi-.

Thus in order to protect ourselves from undesirable development on a do-as-you-please basis, we agree to conform to certain types and magnitudes of garden development and to inform the local authority of our intentions.

There is no need to imagine that there will be difficulties in obtaining the go-ahead to build our little workshop. The basic principle adopted by local planning authorities is that householders should be free to carry out whatever development they wish,

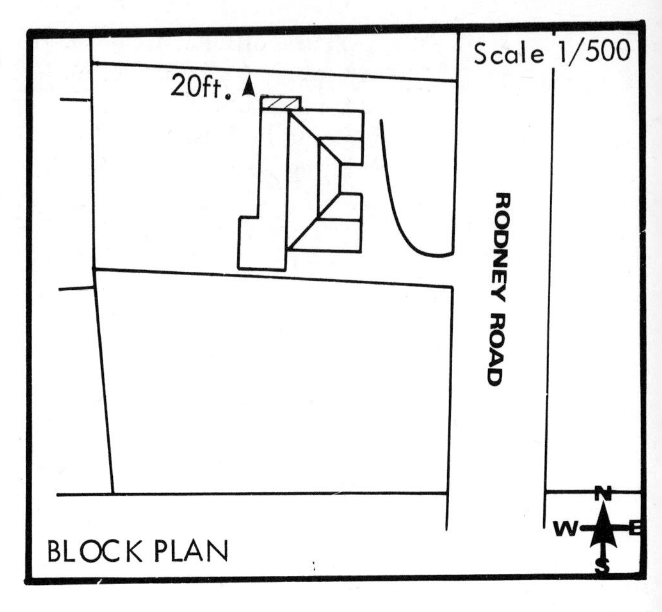

Fig. 13/1(a) Typical block plan with position of workshop shown shaded. The important feature is the distance indicated from the neighbour's boundary.

unless there is good planning reason for refusing permission. Even then, the authority will discuss matters with you to suggest ways and means of effecting a solution, perhaps by resiting the building or using a different finish to conform with local tradition.

So with some idea in our mind of what we want – perhaps we have already made a sketch of our anticipated workshop – we pay a visit to the local authority offices. There we ask them for the necessary forms to apply for permission to build a small hobby workshop in the garden.

If you make a telephone call first, it may be actually possible to see someone in the Planning Office. This is very useful because in a few minutes you can explain what you want to do and he can immediately point out any possible snags. In effect you are picking his brains so that when your application goes before the committee there is no reason for it being temporarily held up due to some slight unforeseen technicality, such as an unacceptable colour for the roof, for example.

The first difficulty may perhaps arise in your mind when you are asked whether you are applying only under "Building Regulations" or whether you need "Planning Permission" as well. Let us deal with the "Building Regulations" application first.

This is a simple application which is required for every type of development (yours will probably be considered as "Minor Building Works" – !). The form normally simply requires your name, address and brief details of the proposed building. Accompanying this application form you must also enclose two copies of a detailed drawing showing a plan, elevations and a section through the proposed building on which is shown constructional details such as timber sizes, position of damp-proof course, type of roof covering, drainage, foundations, etc. This has

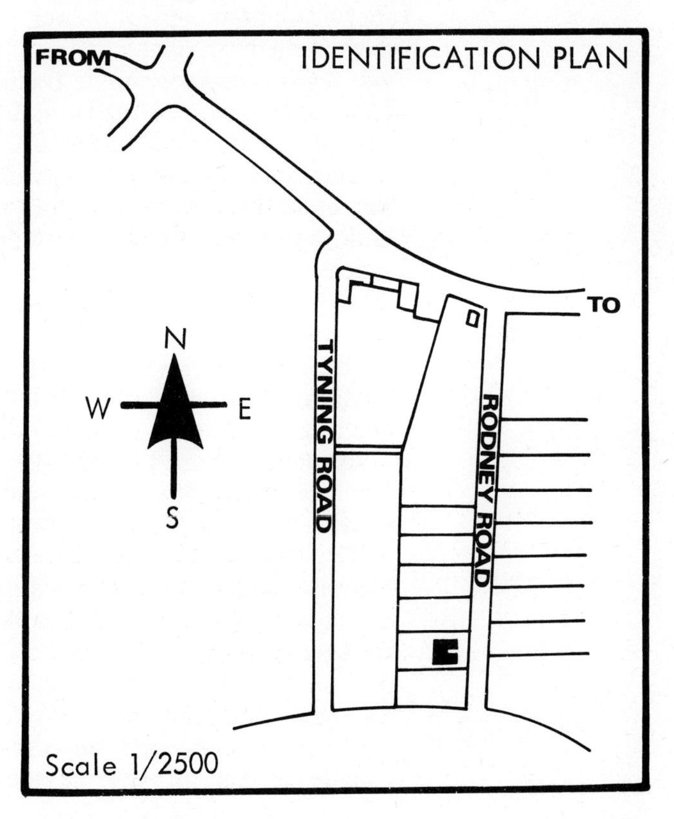

Fig. 13/1(b) Traced from local Authority map, the plan identifies the site for the Building Inspector.

to be drawn to a scale of not less than 1 : 100 or one inch to equal eight feet if that is more convenient.

If you are buying a sectional workshop from one of the firms advertising in various publications, they will probably supply you with the necessary drawings as part of their service.

Also, two copies of a plan showing the whole of your property and the location of the proposed workshop (usually shaded red), in relation to the houses and roads adjoining. This Block Plan is to be to a scale of not less than 1 : 200, that is approximately 1/16 inch represents one foot.

To identify your site accurately at the Planning Office, a Site, Key or Identification Plan may be asked for. This will be to a scale of 1 : 2500, and a word to the Planning Office will probably produce for you a photo-copy of the relevant portion of their master site plan of the district. Alternatively, they will let you make a simple tracing. On both Block and Site Plans show North to assist in identifying and state the scale on all drawings.

This is all you will probably need to obtain sanction under Building Regulations to proceed with the building of the long awaited workshop. In fact the process of making this application is, to my mind, a good thing, as it makes us think carefully about the siting of our workshop, its construction and appearance. We know that our effort, even if only "Minor Building Works" will be an asset to and enhance our property.

But what about "Planning Permission"? In those situations when it is needed, how do we go about obtaining it?

The paperwork necessary is simply the completion of a few more forms, on which many of the questions will not be applicable to your circumstances, such as "how many staff you intend to employ on the site" and "how trade effluents and refuse will be disposed of" – !

More drawings will be needed, but these are usually repeats of those already required for the

Building Regulations Application. So beyond a visit to the local post office, public library or other location of a photo-copying machine, there is little that need cause any trouble.

The information given above refers to the particular district in which the writer lives and is based on such an experience just successfully completed; other local authorities may differ slightly in their requirements. Enquiry to the Planning Office will immediately tell you if Planning Permission is necessary or whether Building Regulations Approval is all that is required.

Do the drawings in black ink; a hard, fine, fibre tipped pen is ideal. As photo-copying is cheap, make an extra copy for yourself; it is sure to come in useful at a later date.

If you rent your property, the owner must be informed of your intentions; also mention to your neighbours what you intend doing. They can object, just as you can to their garden developments, but it is for the planning authority to decide how much importance, if any, to attach to objections and it is the planning authority who take the decision.

Certain types of development have also to be advertised in specified manners – usually either by attaching a notice to the property to be developed or by an advertisement in a local newspaper. These are mainly developments affecting a listed building or a Conservation Area or what is normally described as "bad neighbour" development. This is, as its name suggests, development which is likely to affect adversely neighbouring properties. Within this category come such things as building over 20 metres high, sewage works, a slaughterhouse etc., so it is not likely to trouble you when building your home workshop but in any event the planning office will tell you when you need to do this as they cannot consider the application until it has been done.

So out with the drawing board, start planning that workshop and may you spend many happy and fruitful hours in it.

14. FIXING THINGS TOGETHER

"How shall I fix it together?" is a question often asked by the newcomer, setting up his own workshop for the first time. With experience, the choice of method or technique becomes obvious, but at the beginning it is very easy to make difficulties or even ruin the job if the incorrect choice is made.

It is rarely that the inexperienced will choose a totally wrong technique. For example, he is unlikely to try and nail two pieces of steel together, or solder a couple of bits of wood! But in the number of options available, he may find some difficulty in choosing the best from those that will just "do".

In the home workshop both wood and metal will require to be joined, both to themselves and to each other, and as our first efforts in setting up our workshop will necessitate working in wood, let us start with that material.

Wood may be fixed together by nails, screws and glue. The first of these, *nails*, come in three main types. Wire nails, with round shanks and flat heads, used for ordinary carpentry work where the visible head does not matter. Oval nails, with oval shanks and small heads that may be punched below the surface. Suitable for finer work where the heads are to be unobtrusive. And panel pins, which are small and have thin round shanks. These are for the most delicate work and may have a pointed head over which the wood closes once they have been driven home.

Nails are suitable in situations where location is

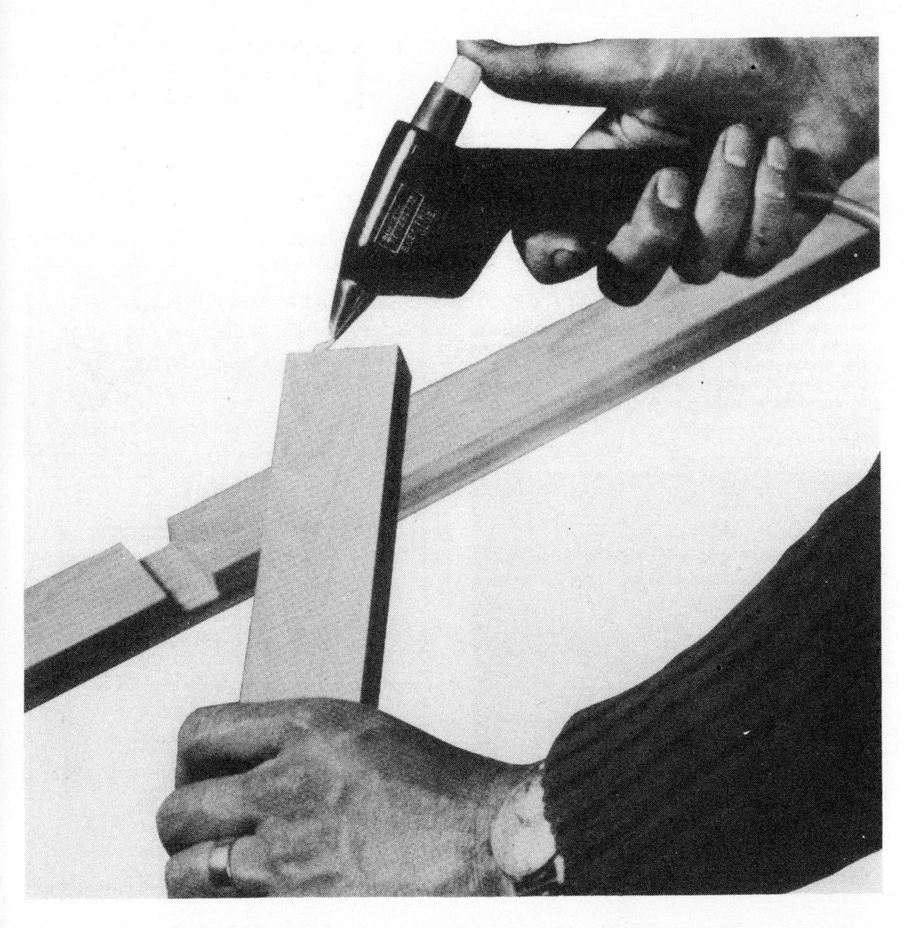

Fig. 14/1 Adhesive or sealer application by electrically heated "gun" using "sticks" of glue.

the main requirement, rather than strength. Remember that a nail holds by the wood fibres pressing against the shank of the nail, thus the holding power is purely frictional and a pull that is sufficient to overcome this friction will withdraw the nail. Thus to obtain maximum strength in a nailed joint, insert the nails at right angles to the direction of the pull trying to separate the joint. The simplest and typical example is how we drive in a nail to hang a picture.

Screws come in a multitude of diameters and lengths, with countersunk and round heads and in a variety of metals. They grip by cutting into the wood so as to form a nut which grips the screw thread tightly. Although we can generally drive the screw straight into softwood with a screwdriver, we will find it necessary in hardwood to drill a pilot hole for the thread and a clearance hole for the screw shank. If using a countersunk head screw, a suitable depression drilled with a rose bit (or end of a twist drill), will finish off the job neatly.

Fixing screws will be found to be easy if we use the correct size of screwdriver and make certain that its end is carefully ground or filed to fit the slot in the screw head.

Try not to use steel nails or screws in oak as the tannic acid present in the oak will react with the metal to form a dark, inky like stain, spoiling the appearance of the wood. Eventually the nails or screws will be eaten away by this corrosive attack.

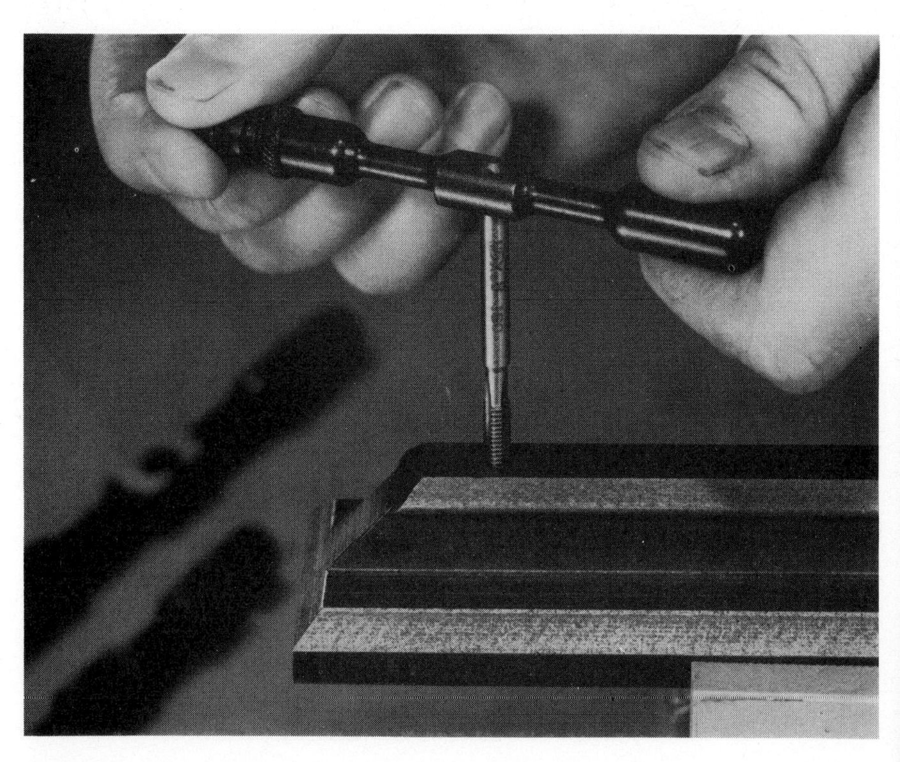

Fig. 14/2 The ability to cut threads in metals is a valuable development in workshop activities. This picture shows the tapping of a 5 mm thread in a cast iron block.

Wood may be bolted together, but as details of this technique have been commented upon in the chapter on bench making, we will allow that to suffice.

Adhesives have come a long way since the old-fashioned fish and animal glues of my early working days. I remember being told by an old cabinetmaker that the perfect way to treat a cut finger was to dip it in the hot glue-pot. Thereafter I was more petrified of the likely cure than the cut!

The joining power of such glues depended on their ability to penetrate the surfaces being joined and forming minute pegs or filaments which mechanically held them together. In general this required that the glue be heated to increase its fluidity and penetrating power; such glues were only suitable for use on surfaces of an absorbent nature. Many modern adhesives however, have the ability to join together the non-absorbent surfaces of metals, glass and plastics. Their linking power is dependent on the ability of the molecules of the adhesive to form a cohesive bond with the surface molecules of the work. For this reason we find a considerable expansion in the range of adhesives available, and, rather than wasting space listing them, would advise the reader to read carefully the instructions the next time he buys a tube of glue and note the materials that may and may not be joined with it.

Even glue application has changed with the years and we now find an electrically heated "gun" being used to make the operation fast, easy and clean. The adhesive or sealer is supplied in stick form and is usually based on a thermoplastic compound which melts at around 150°C.

Having said that metals may be bonded with the new adhesives we can now proceed to consider other ways of joining metal parts. They divide themselves naturally into mechanical methods and those requiring the application of heat.

Of the mechanical means, screws or bolts together with nuts are probably the most commonly employed. Screws are normally threaded right up to the

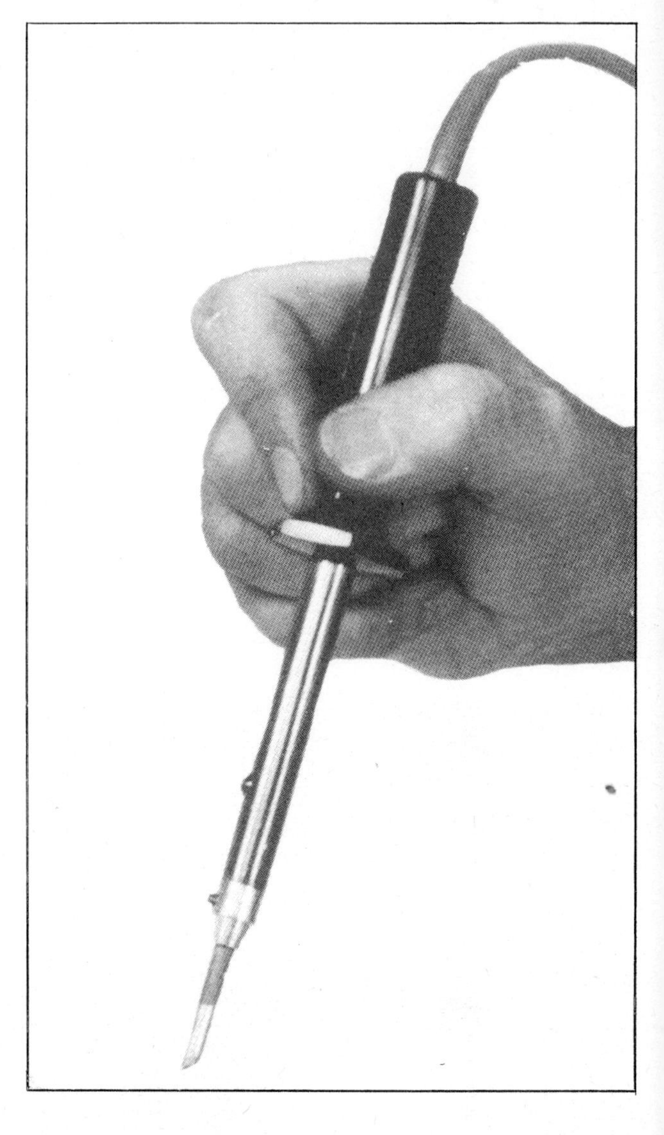

Fig. 14/3 Soft soldering is one of the most useful of metal-joining processes. The 25 watt electric soldering bit shown is ideal for the home workshop, where small model making or radio building is carried out.

head, while bolts have a plain portion, being generally threaded for about one-third of their shank length. Heads may be of many forms, round, countersunk and hexagonal being the most common.

The screw or bolt is locked with a nut which is almost certain to be hexagonal – although, just to be different, the coach bolts used on timber structures, always have square nuts!

In place of the nut we may fit the screw or bolt into a threaded hole in the work. The process of producing this threaded hole requires us to drill an undersize hole and cut therein a thread with a tool called a "tap". Obviously the "tap" used (see Fig 14.2) needs to be of the same diameter and thread size as the screw or bolt, so it could be an expensive business stocking our workshop with every size of tap we might need. However, if we buy this equipment only as we require it, and stick to the common sizes, we will not find the cost excessive and over the years will build up a valuable range of thread sizes.

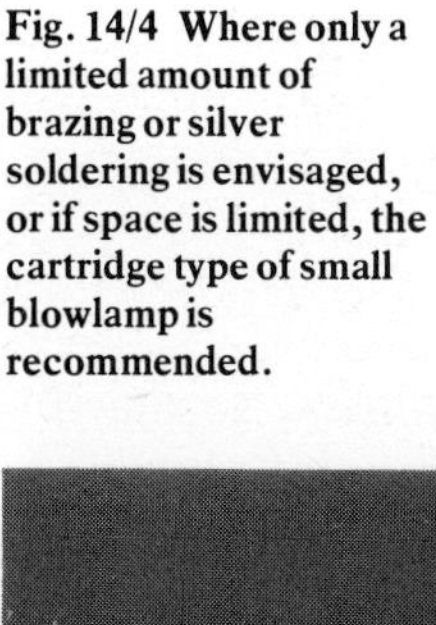

Fig. 14/4 Where only a limited amount of brazing or silver soldering is envisaged, or if space is limited, the cartridge type of small blowlamp is recommended.

If we can afford it, it is worth purchasing the thread "die" at the same time as getting the taps. This device cuts the external thread on a piece of rod of correct diameter so making the "screw" or "bolt".

The other way of fixing metal parts together mechanically is by riveting. This is a much older

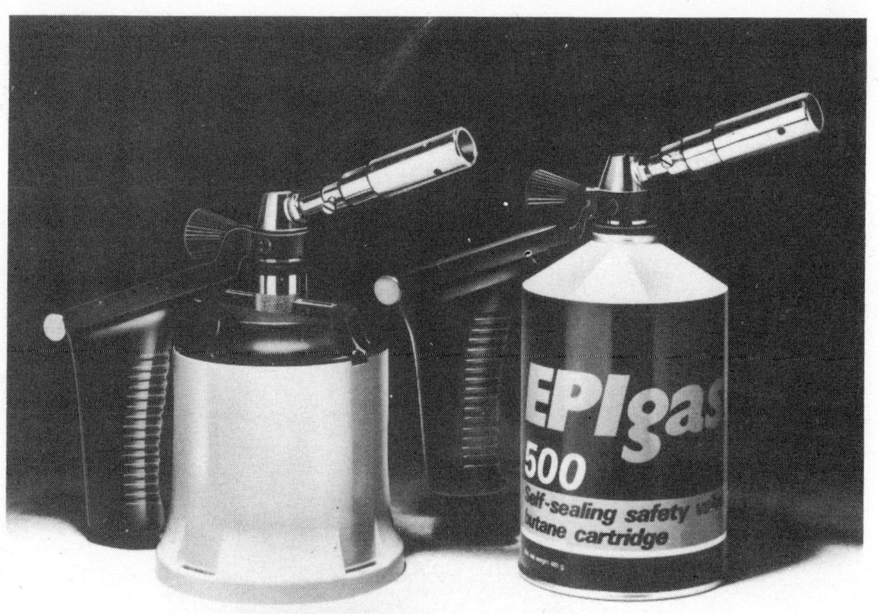

technique than screwing and has been practised ever since metals were first used. Screwing parts together may be looked upon as temporary, but riveting is a more permanent fixing process.

Basically a rod of metal, iron, copper, brass, aluminium, is passed through close fitting holes in the parts to be joined. The ball end of an engineer's hammer is then used to form a head by beating over the protruding ends of the rod. If very large rivets are being used they are raised to red heat prior to being inserted. This makes the heading operation easier and also, as the rivet contracts on cooling, the parts being joined are drawn more tightly together.

A number of methods exist by which metal parts may be joined to each other by the application of a heat source. The simplest of these is soft soldering in which the joining medium is a low melting point alloy of lead and tin.

Most of the common metals can be soft soldered, although aluminium requires rather special techniques. The process needs cleanliness, so the surfaces to be joined have to be thoroughly cleaned by filing, use of emery cloth or wire wool. To ensure that cleanliness is maintained during heating, a flux is used to cover and protect the metal surfaces.

Because soft solder melts at a comparatively low

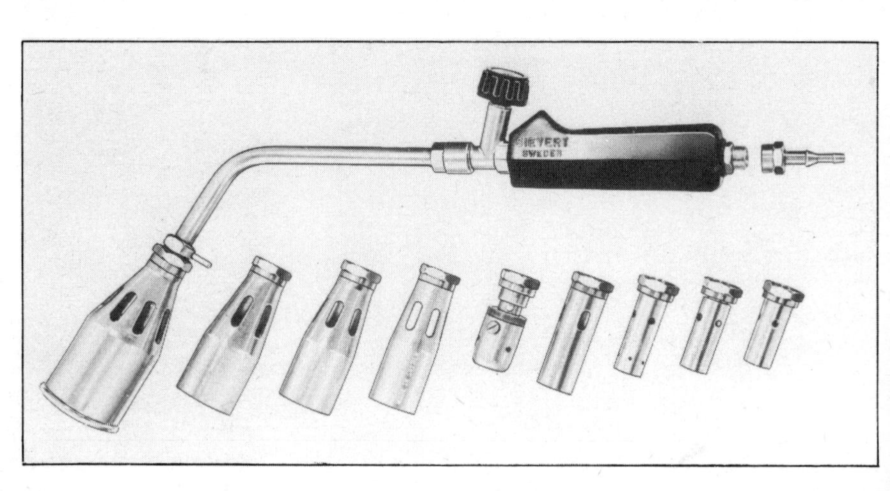

Fig. 14/5 A wide range of nozzles is available for liquified gas equipment, permitting from the smallest to very large heating requirements.

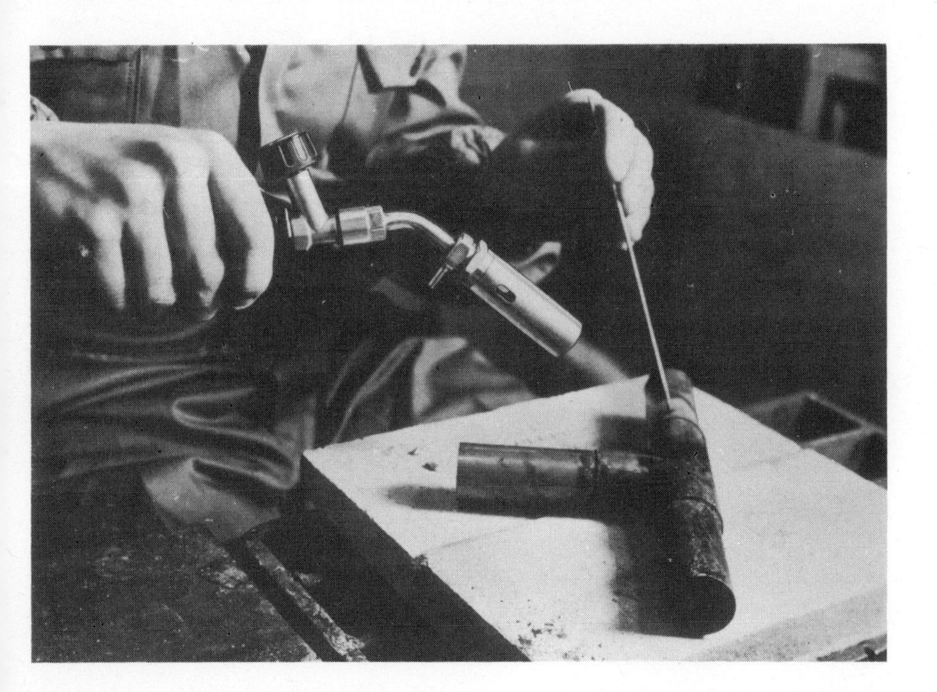

Fig. 14/6 The availability of heating equipment so that brazing operations can be carried out greatly increases the scope of work. With the construction of a simple hearth, various bending and forging operations may be effectively performed.

temperature, around 200°C., heat may be applied from a copper headed soldering "bit", the size of which – and hence its heat content – depending on the size of the joint to be soldered.

The principle of the soldering operation is that the solder melts and spreads over, "wetting" the clean, hot, metal surface, with which it combines or alloys. By using a more powerful source of heat, such as a blow-torch flame, we can carry out a similar soldering process but this time using brass as the joining agent instead of the soft, lead/tin, solder. We call this new technique, "hard soldering" or "brazing" and, although needing a considerably higher temperature (800 – 900°C), it forms a very strong joint indeed. The requirements of cleanliness and use of the correct flux are as before.

Like soft solder, brass is an alloy, this time of copper and zinc. If some silver is added to the brass, the melting point of the resulting alloy falls appreciably and we can make a series of hard solders

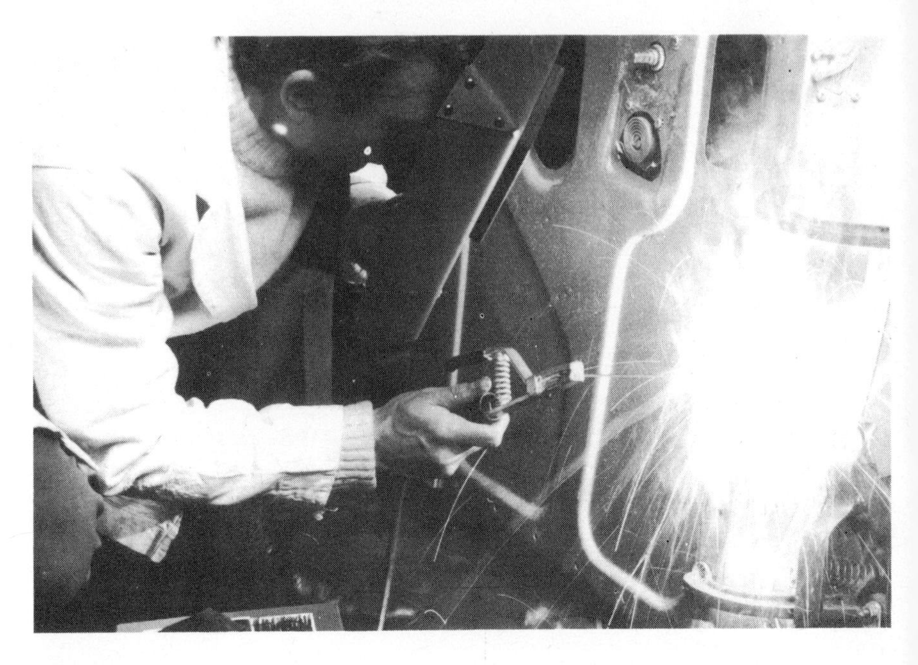

Fig. 14/7 Arc-welding equipment operating from the normal domestic supply may appeal to the home workshopper interested in light metalworking.

(silver solders) with melting points down to as low as just over 600°C. Needless to say, the silver content makes them rather expensive! However, if used sparingly for precise jobs they make a beautifully clean and neat joint.

A different principle is used when welding metals together. Here the two pieces of metal are heated to such a temperature that they themselves melt and run together forming a completely homogeneous joint, with or without the addition of extra metal from a filler rod.

Two main processes are used to effect such a joint, using, in the first place, acetylene and oxygen gases from two cylinders, which mix in a blowpipe and burn at its nozzle, with a small but intensely hot flame. This equipment, although useful, is not really suitable for the home workshop because of the high rental charges on the gas cylinders. Lately a set with very small cylinders has been made available, but does not yet seem to be as popular as one might have supposed.

The alternative method of welding, particularly of steel, is that using an electric arc. The arc is struck between the work and a consumable, flux-coated electrode fixed in a handpiece. The low voltage, high amperage current from a transformer instantly melts the parent metal forming a molten pool which, with practice, may be controlled and caused to move along the joint.

Many small arc welders of up to 140 amperes output are available which work from the normal domestic 13 amp mains supply. From my own experience I find such an apparatus an almost indispensable piece of workshop equipment.

There is obviously much more to all these processes than I have been able to describe in this brief chapter, but, having "broken the ice" you can now refer to the many relevant Argus publications for the detailed techniques of carrying out these very useful methods.

Fig. 15/1 Cowells Miniature Drilling Machine.

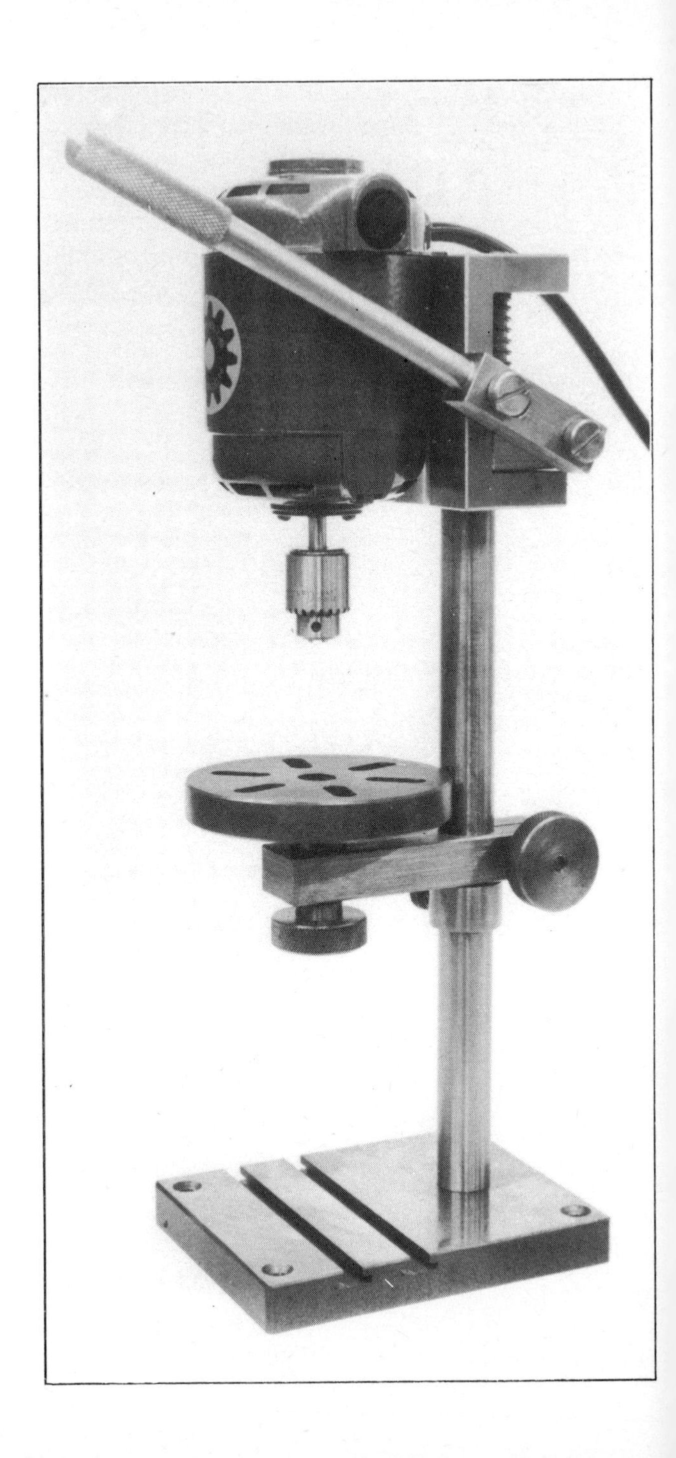

15. THE COMPACT WORKSHOP

Earlier in this book we made the comment that the pleasure to be gained from working at a craft was not dependent on the size of work attempted. Similarly, limited space is no deterrent to owning a well equipped machine shop if our interests lie with any of the many different model engineering pursuits.

Many owners of garden shed workshops may also feel that the cold winter nights put a stop to the evenings they would normally spend in their workshop.

To all readers who recognise these problems, I would commend the idea of developing a compact workshop which will occupy little space and may even be completely packed away so as to be unobtrusive after use.

For some time now I have had the opportunity to study and use the machine tools suitable for such a workshop, manufactured by Cowell Engineering Ltd., of Norwich, who advertise regularly in the technical hobby magazines, and it is of this experience that I now write.

The model builder will probably find that his first need is to be able precisely to drill holes in metal parts. The miniature drilling machine in Fig 15, has a chuck capacity of from No 80 (0.344 mm/0.0135 in.) to 5/32 in. (3.97 mm) diameter. The machine is accurately made and the fully adjustable table is identical to the faceplate used on the Cowells 90 lathe. It may also therefore be interchanged with the latter's three- and four-jaw chucks. The motor runs,

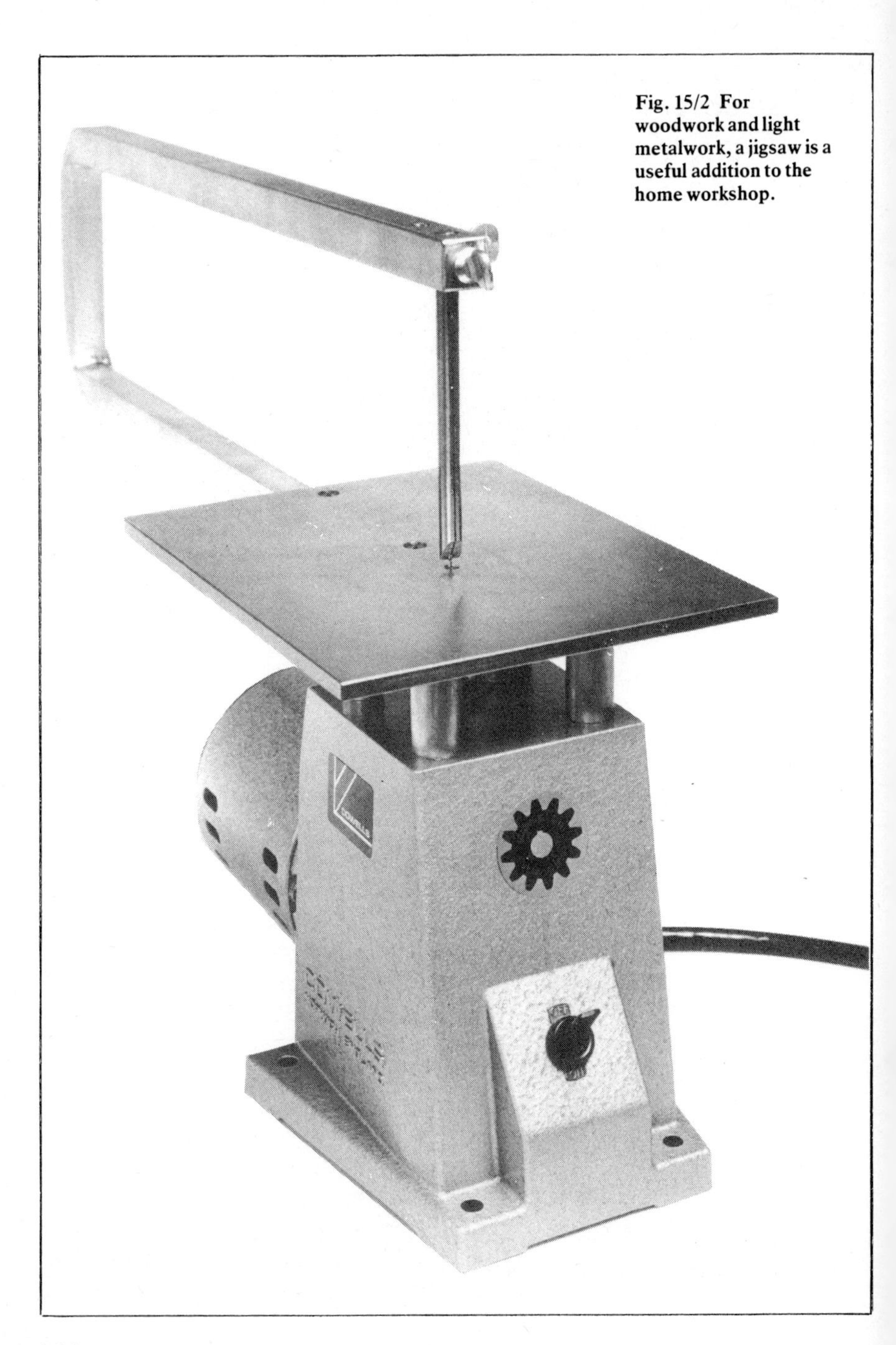

Fig. 15/2 For woodwork and light metalwork, a jigsaw is a useful addition to the home workshop.

Fig. 15/3 The new Cowells 90 Lathe, the perfect machine tool for the compact workshop.

off load, at 10,000 rpm but, if desired, a variable speed foot control may be fitted.

Still thinking about the model builder, the jigsaw is a sound design with built-in safety features. A full range of blades is available for cutting wood, plastics and metal.

For the model engineer requiring the facilities to do more advanced machining operations, the Cowells 90 lathe is a substantially made machine with all the precision and facilities normally featured in a professional engineer's lathe. A very full range of accessories is available including rear toolpost, chucks, indexing unit, saw table, machine vice and vertical slide, the last being almost a required item in any miniature engineer's list of equipment. The new version of the 90 lathe is illustrated.

Mini-Machine Shop

Some of the illustrations show work being carried out in the author's own workshop where a Cowells lathe was used to machine the parts for a Stuart

Fig. 15/4 Boring a model steam engine cylinder on the faceplace of the Cowells 90 Lathe.

Fig. 15/5 Work in the self-centring chuck of the Cowells lathe.

Fig. 15/6 The circular saw attachment is equally suited to cutting metal – here 16g. brass sheet is being cut – or fine work in wood.

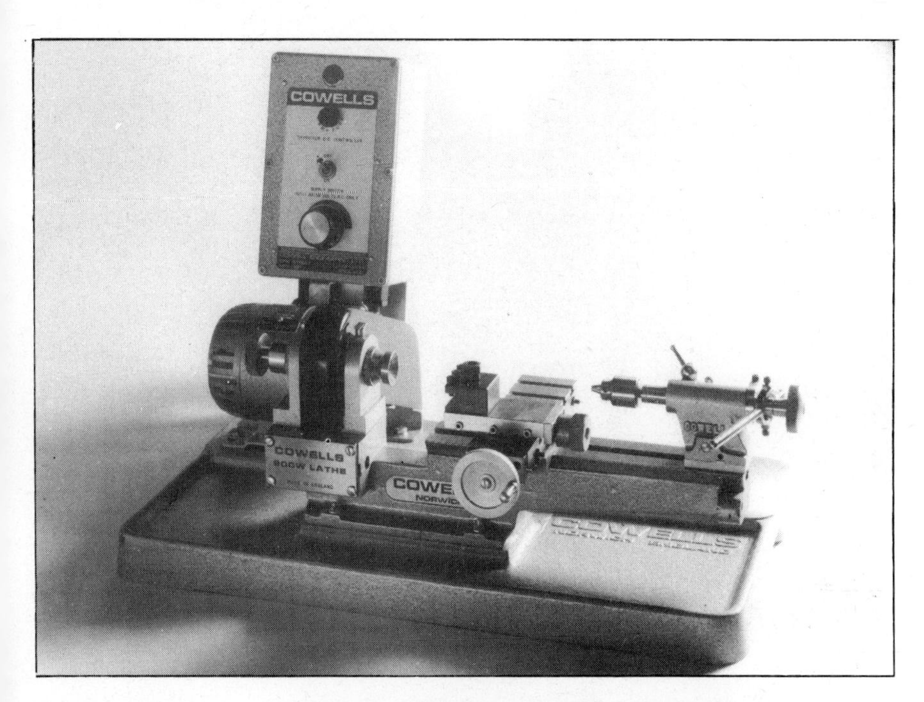

Fig. 15/7 Special versions of the Cowells 90 lathe are available; this is the 90 CW for horological enthusiasts both amateur and professional.

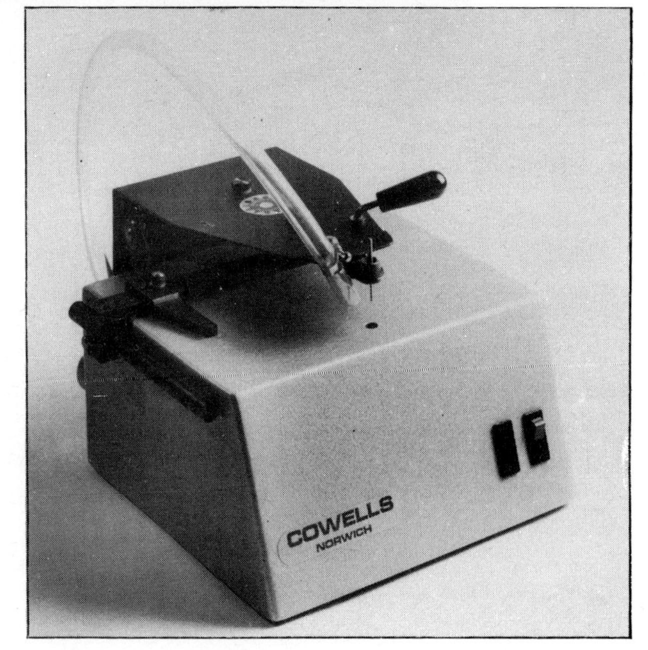

Fig. 15/8 The electronics expert will find this unusual type of drilling machine for printed circuit boards will meet his requirements.

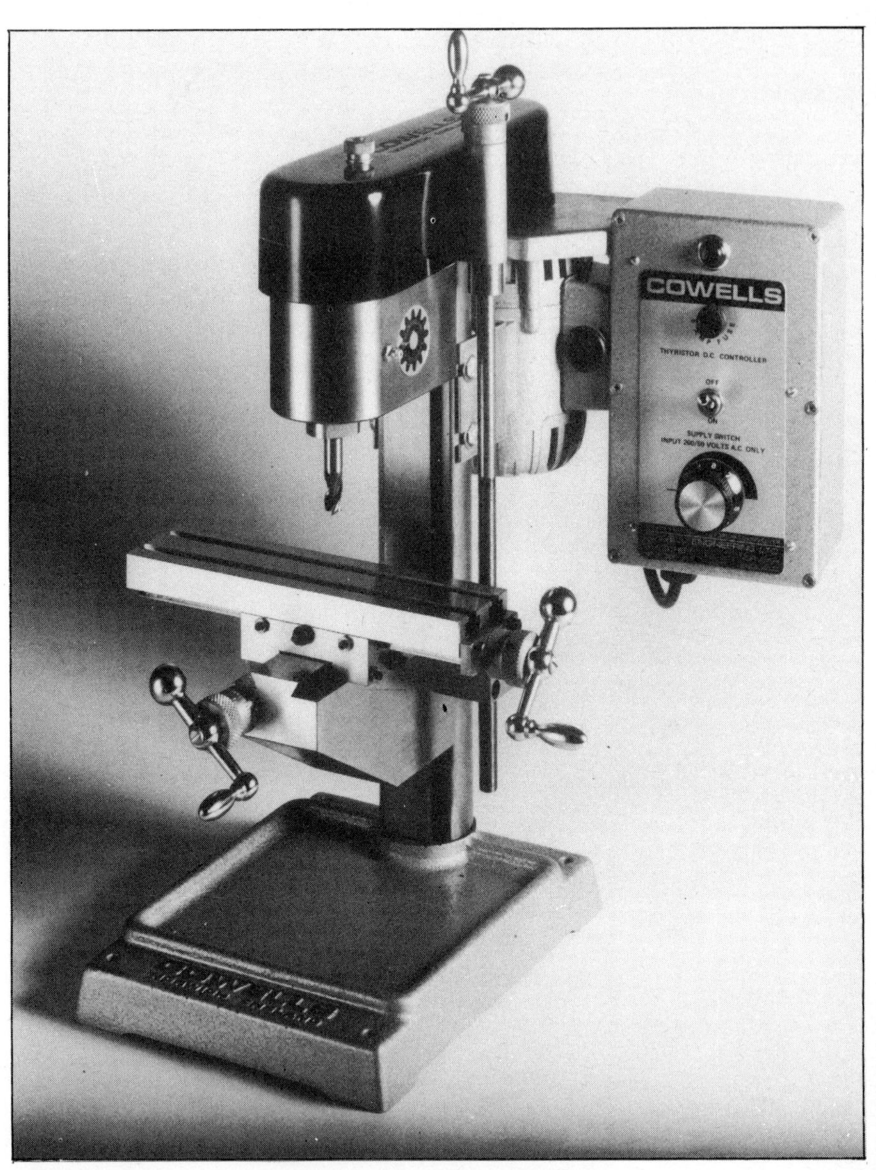

Fig. 15/9 Vertical Milling Machine.

Turner twin-cylinder mill engine and a $3\frac{1}{2}$ in. gauge model steam locomotive. A handbook for the Cowells 90 lathe is available which fully details and illustrates the work that may be carried out on this excellent machine.

A natural partner for the lathe is the vertical milling machine designed to use many of the accessories which are supplied for the lathe. The motor is thyristor controlled giving an infinitely variable speed range up to 3,000 rpm. End mills up to 10 mm ($^{3}/_{8}$ in.) and flycutters up to 50 mm (2 inches) diameter may be accommodated.

With this equipment there is little machine work that will not be within the capabilities of the owner. No more than an area of two feet square is needed to fit in the lathe, milling machine and drilling machine for model engineering. Alternatively the machines may be fitted in a wardrobe or cupboard and shut away from view when not in use.